The Patrick Moore Practical Astronomy Series

W0037141

More information about this series at http://www.springer.com/series/3192

Finding a Million-Star Hotel

An Astro-Tourist's Guide to Dark Sky Places

Bob Mizon

 Springer

Bob Mizon
Wimborne, Dorset, UK

ISSN 1431-9756   ISSN 2197-6562   (electronic)
The Patrick Moore Practical Astronomy Series
ISBN 978-3-319-33854-5   ISBN 978-3-319-33855-2   (eBook)
DOI 10.1007/978-3-319-33855-2

Library of Congress Control Number: 2016942567

Printed on acid-free paper

This Springer imprint is published by Springer Nature
The registered company is Springer International Publishing AG Switzerland

Stargazing gives us access to orders
of events, and scales of time and space, which
are beyond our capacity to imagine.

–Robert Macfarlane: The Wild Places,
Granta Publications, 2007

Preface

London's Underground may not be the most suitable place for contemplating the universe, but on a journey in the late 1990s from Green Park Underground station to London Waterloo, my eye was caught by a car card, one of those small advertisements that run along the carriages below the roof line. Unusually, it advertised stargazing holidays. Its message: that jaded Londoners could seek rest and relaxation beneath the star-strewn night skies of Pembrokeshire, southwest Wales, away from the bustle and the light pollution of the big city.

I spent my childhood in East London in the 1950s, a time of post-war rationing and austerity. Road lights were rather sparse and fairly dim and went off in the late evening. During the winter, when the nights were long and I could play outside with friends before bedtime, I noticed the stars and planets in the sky from the relative darkness between the lights. My interest in astronomy was sparked. When I dared to stay out late, risking my mother's wrath, I would wait for the lights to go off and see just what the universe had to offer, marveling at its shifting lights and star patterns, before going home to face the music.

Then came the decades of intensive lighting, most of it poorly directed, shining outwards and upwards. On all night, it erased the starry scene that had formed part of my childhood.

Now, in the twenty-first century, the vast majority of urban children have had no experience of the real night sky. For many, their evening world has shrunk to the size of a small screen or two in a bedroom at home. Of course, they can use their computers to call up, should they wish to, images of startling beauty taken by today's giant telescopes or by amateur astronomers with the latest cameras and image processing programs. Here is a sad irony of modern life. Most people in the developed world, the town and city dwellers, have lost sight of the near and far

Fig. 1 "Half the park is after dark." (Photo courtesy Dr. Tyler Nordgren)

universe from their gardens and parks because so much lighting now shines where it is not needed, while other technologies, those of spacecraft and state-of-the-art optics, can show them in their own homes what the universe looks like, on televisions, tablets, laptops, and telephones. The immediacy and reality of the universe have receded into the pixels.

Perhaps we have never really lost our desire to see for ourselves what's up there. Recently, astronomy clubs and courses have seen a marked increase in attendance.

Outdoor observing sessions attract ever greater numbers of hopeful stargazers. Teachers know that, of all their science topics, space is the one that is likely to hold the attention of their classes. People are seeing more stars in cities, thanks to the work of campaigners for better directed lighting, and the lighting industry has responded over the years with increasing numbers of downward-directed lamps on our streets.

The trickle of astro-tourists from urban areas has become a stream, though not yet a torrent. Many hotels and campsites and indeed whole counties, national parks, and nature reserves now advertise and protect their stars. For example, national parks in the United States have published striking posters (Fig. 1) designed by Dr. Tyler Nordgren, telling us that "Half the park is after dark".[1] Like so many other such areas, Maine's Acadia National Park now holds star parties, where volunteer local astronomers and park rangers guide visitors through the constellations and zoom in on planets and other night sky wonders from Cadillac Mountain.

That single Underground car card, with its Welsh stars, nowadays has competition. Ever more numerous ads for breaks beneath the stars appear in newspapers, on the Internet, in travel magazines, and cruise line leaflets. This book will examine a new tide of discovery, as urbanites reconnect with their ancient heritage above. *Finding a Million-Star Hotel* is aimed primarily at those with limited possibilities for astronomy, who seek out the thrill of the stars they cannot see at home. It will also look to a future when the real night sky will be readily accessible even from urban areas, just as images of the Crab Nebula and the Andromeda Galaxy can nowadays be revealed by a quick web search.

Wimborne, UK   Bob Mizon

---

[1] www.ecology.com/2013/05/10/national-park-service-posters.

Acknowledgments

For all their help and encouragement in preparing this book, I thank numerous members of the British Astronomical Association and its Commission for Dark Skies; all the generous and talented astrophotographers and amateur astronomers in the United Kingdom and the United States who supplied images and related their experiences; the tourist offices, national parks, and hotel and campsite owners who answered inquiries and sent material; and the dedicated staff of Southmead Hospital in Bristol and the Royal Bournemouth Hospital, without whose skill and dedication this book would certainly never have been written.

My special thanks to my wife Pam for her enthusiastic support.

Contents

About the Author

Bob Mizon, MBE, FRAS [Bob likes to point out that his name rhymes with "horizon"], is a graduate in modern languages, but is much better known as an astronomer. Having taught for 26 years at Poole Grammar School—where he met his wife Pam—he embarked on a rather daring career change in 1996. Responding to a lifelong love of astronomy, he became a planetarium operator and now takes a stunning mobile dome into schools, youth groups, and societies all over southern England. Over 130,000 people have experienced a tour of the universe with Bob at the controls!

Bob is best known in the scientific and environmental community as the coordinator of the British Astronomical Association's Campaign for Dark Skies, which aims to turn back the tide of light pollution that has seriously affected our view of the stars over the last fifty years. Glare, light intrusion, and skyglow have become the norm nowadays, a situation hardly compatible with a society that is supposed to be saving energy and protecting the environment.

Chapter 1

What Are Dark Sky Places?

> The sight of the stars makes me dream.
>
> –Vincent Van Gogh

One day in August several years ago, my wife Pam and I were staying in a hotel in north Norfolk, one of the few really dark areas left in England. There were meteors due: the annual Perseid stream. I told the owner this, but he already knew all about them. Guests often went out to see the stars, he said, and when they did he would switch off the exterior light, already quite muted and well angled downwards, and darkness would reign.

We enjoyed the flashing Perseids on that warm night as we looked out across the North Sea, where only the tiny constellation of a far-off ship's lights could be seen beneath what seemed a million stars (Fig. 1.1).

If only everyone could step outside and see such things…

There are many who are trying to make that possible.

In November 2011, the British media widely reported the creation of Europe's first International Dark-Sky Reserve. 267 sq. miles (692 sq. km) of the scenic Exmoor National Park, in the southwest of England, had been awarded this status, with the International Dark-Sky Association's silver-tier ranking. The IDA awards are recognized worldwide. The night sky over Exmoor was thus confirmed as a special resource worthy of protection, and the star-strewn darkness above it, riven by the

© Springer International Publishing Switzerland 2016   1
B. Mizon, *Finding a Million-Star Hotel*, The Patrick Moore Practical Astronomy Series,
DOI 10.1007/978-3-319-33855-2_1

Fig. 1.1 Summer stars and the Milky Way over the sea. (Photo: James Hilder)

silver stream of the Milky Way, is now actively maintained. Controls are in place to prevent wasted artificial light tainting its night skies.

The pre-submission process for IDA award bids is not a simple one. It requires the organizational skills and patience of a dedicated team. In the case of Exmoor, dark sky measurements and photography were carried out by local individuals and national bodies. Exmoor photographer David Brabban toured the park at night, metering sky darkness and capturing its breathtaking night panoramas (Fig. 1.2). Astronomer and dark sky advisor Steve Owens and lighting consultant Jim Paterson, a firm friend of the night sky, weighed in with their expertise, as they have in other places seeking dark sky status. Members of the British Astronomical Association's Campaign (now Commission) for Dark Skies carried out sky measurements with calibrated Sky Quality Meters (SQMs), on one occasion dodging the hooves of inquisitive local cattle. A lighting audit was organized. Local organizations and householders entered into agreements on good-quality lighting at their premises. Evidence of dark skies for the IDA also included supporting letters from local and national bodies and prominent people backing the dark sky status bid. Local administrations and businesses, astronomical societies, schools, nature groups and National Park appreciation societies were actively involved.

In May 2015, the IDA announced that Pickett State Park and Pogue Creek Canyon State Natural Area, at the edge of the Cumberland Plateau in northern Tennessee, had joined the dark sky club and had become a silver-tier International

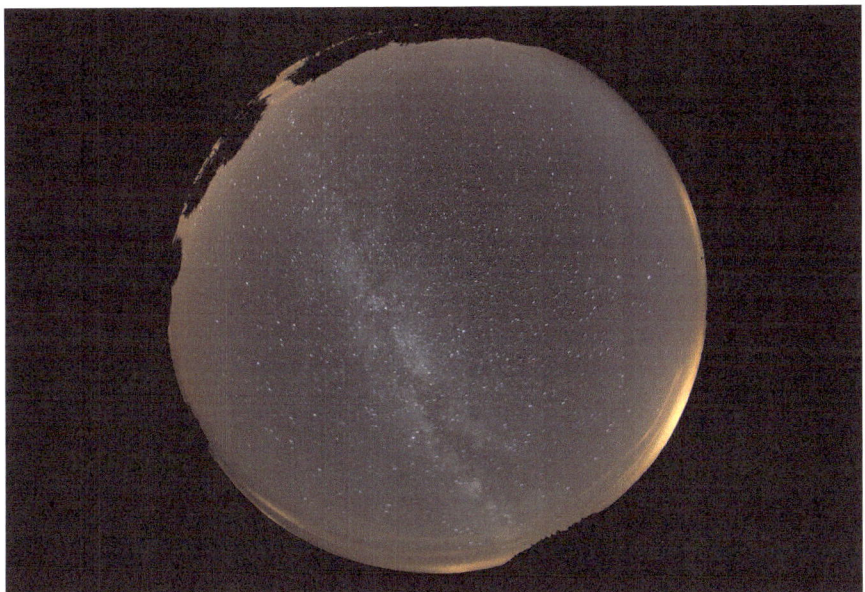

Fig. 1.2 The Milky Way arches over Exmoor from Perseus (*top*) to Sagittarius. (Photo: David Brabban)

Dark-Sky Park within its award scheme. The Pickett-Pogue International Dark-Sky Park worked with local communities and administrators, bringing local lighting to star-friendly standards. The IDA reported that "in partnership with the Barnard-Seyfert Astronomical Society and the Space Science Outreach at the University of Tennessee, the park has developed a strong interpretive program. Activities include an annual New Year's Eve hike, a Junior Ranger camp each July, an annual star party, and regular dark skies programming during peak visitor season in the summer" (Fig. 1.3).

Projects such as the Exmoor and Pickett-Pogue star reserves remind us of the need to defend sites with dark night skies, and affirm the value and importance of a view of the cosmos, now lost to so many. Both humans and wildlife need the night; most of the world's species are nocturnal, as it's a pretty obvious predator avoidance strategy. And the night sky needs our intervention, to restore it to something like its old self, for all to enjoy.

International Dark-Sky Places (IDSP, listed in Appendix 1) increase in number every year, and must offer and facilitate public access to the night sky within their boundaries, creating programs to encourage and educate visitors. They must also improve and preserve night sky vistas by ensuring and maintaining 'star-quality' lighting, not only in the immediate area of observing sites but at more distant locations, since light travels!

Star trails over Pickett-Pogue Dark Sky Park
July 2015 T Wellington

Fig. 1.3 Star trails over the Pickett-Pogue International Dark Sky Park. (Photo: Theo Wellington)

Natural Bridges National Monument (Utah), Cherry Springs State Park (Pennsylvania) and Clayton Lake State Park (New Mexico) were the first American dark sky reserves, in 2006, 2008 and 2010, respectively. They have since been joined by several others (www.darksky.org/night-sky-conservation/dark-sky-parks).

Galloway Forest Park in southwest Scotland, the South Downs National Park near the south coast of the UK, the Brecon Beacons National Park in Wales, Northumberland International Dark Sky Park/Kielder Water Forest Park, the island of Sark in the English Channel and other starry places within the British Isles have secured, or are working towards, IDSP status. The UK, at the time of writing in 2016, has more official IDA Dark Sky Places than any other country in the world outside the United States (Fig. 1.4). (www.britastro.org/dark-skies/bestukastroloca-tionmap1.html)

The benefits of dark sky protection schemes are evident. If, as already mentioned, increased numbers of visitors are attracted to areas of guaranteed darkness in the late autumn/fall, winter and early spring off-seasons for tourism, they will find that the night sky is just as glorious as it is in the summer—even more so in the opinion of many astronomers, and especially in the winter, with its awe-inspiring patterns of remote lights. Orion and its sparkling retinue of constellations

Fig. 1.4 The winter sky over Galloway Forest Park: Orion and Taurus. (Photo: James Hilder)

(Gemini the Twins, Taurus the Bull, Canis Major the Great Dog and others) dominate crisp and cold southern skies. Whatever the season, the public's enjoyment of the stars is enhanced by designated viewing areas with good facilities, and possibly an observatory, with local astronomers helping out. Good publicity for star parties and policies assuring 'star-quality' on-site lighting of car parks and buildings can add much to the observing experience. In the daytime, exercise and fresh clean rural air are a bonus.

The area's wildlife benefits, as does the environment in general. Better lighting means less light intrusion into habitats and fewer adverse effects on nocturnal flora and fauna (Fig. 1.5). It ensures more intelligent energy use and less pollution all around.

The local economy also benefits. All those hopeful stargazers have to sleep somewhere, eat, shop and travel, so tourism-based industries, small businesses and retailers welcome their presence. A new astronomy facility, be it an observatory or a place where local astronomers come to share their telescopes (Fig. 1.6), becomes a great educational resource for local schools and encourages such sky-related enterprises as telescope hire, specialist bookshops, and hotels and campsites catering for starwatchers.

Of course, people should have the optimum night sky to enjoy wherever they live, and in many places modern lighting technologies, dimming and switch-off schemes, and well-informed local authorities have ensured that this is happening.

Fig. 1.5 Insects diverted to a floodlight, which also sends half its emissions into the sky. (Photo: Colin Henshaw)

Fig. 1.6 Public-access astronomy: the Wessex Astronomical Society's observatory (*arrowed*) and astronomy center near Swanage, Dorset. (Photo courtesy WAS)

Skies have become darker in many areas. Those who have been to dark areas to see pristine, protected night skies may well spread the word locally on their return home, recommending good lighting practices where they live, bringing back the stars. Children who are required to learn about the universe in school curricula then have a better chance to see more of it for themselves from their backyards.

The non-profit IDA, based in Tucson, Arizona, was founded in 1988 by professional astronomer Dr. David Crawford and physician and amateur astronomer Dr. Tim Hunter (Fig. 1.7). Its mission: "to preserve and protect the night time environment and our heritage of dark skies through quality outdoor lighting." To access its information sheets and news updates, see www.darksky.org.

The IDA, with the support of other international organizations promoting dark skies, establishes areas where starscapes are improved and protected throughout the world; there is an ever lengthening list of International Dark Sky Communities, International Dark Sky Parks and International Dark Sky Reserves. From the IDA's website:

> The first Community was the town of Flagstaff, Arizona, USA, established in 2001. The first park was the Natural Bridges National Monument, Utah, USA, in 2006. The first Reserve was at Mont-Mégantic, Québec, Canada, in 2008. Since then, a procession of areas

Fig. 1.7 Dr. Tim Hunter (*left*) and Dr. David Crawford, co-founders of the International Dark-Sky Association. (Photo courtesy IDA)

both large and small have gained the IDSP status, in places as far apart as Namibia, Hungary, Ireland and New Zealand. The list grows.

See www.darksky.org/night-sky-conservation/dark sky-places and follow the link…

Preserving dark skies starts locally with a dedicated group of citizens, staff, or volunteers. IDA International Dark Sky Parks and Reserves are home to some of the darkest and most pristine skies in the world. IDA International Dark Sky Communities are filled with concerned citizens working to make their city a little darker. While their skies may not be perfect, they provide examples to the world of cities with responsible outdoor lighting.

Within these areas, there are organized astronomy activities. Local tourist and park administrations create public-access observing sites where visitors can seek out stars with their own binoculars and telescopes, or merely marvel at the universe with unaided eyes. Camping grounds, hostels and hotels put the night sky high on their agendas. A growing number of these have their own dedicated observing areas, instruments that visitors can borrow, and even observatories.

Dark Sky Communities are towns, cities, islands or other small areas dedicated to the protection of the night sky, ensuring best-quality lighting, public education and involvement of the local population in promoting dark skies (Fig. 1.8).

Dark Sky Parks are large-scale parks or publicly accessible areas of similar size with exceptional night-sky views. Their authorities monitor lighting for both star-watching and wildlife benefit, and foster educational and cultural programs. They promote best use of their dark sky resource (Fig. 1.9).

Fig. 1.8 Starry skies over Sark, a Dark Sky Community. (Photo: Martin Morgan-Taylor)

Fig. 1.9 Clear skies over the Elan Valley Dark Sky Park, Wales, promising glorious starscapes later. (Photo: Bob Mizon)

Dark-Sky Reserves are usually larger areas (National Parks, for example) with pristine night skies across much of their territory. They encourage visitors to savor their starry skies and organize outreach programs, events, observing sites and other facilities, for example, observatories. An International Dark Sky Reserve must have a core area satisfying criteria for natural darkness and the quality of the night sky, with a surrounding buffer zone in support (Fig. 1.10).

The three tiers (Gold, Silver, Bronze) within which dark sky status is awarded reflect the visibility of the faintest stars seen from within the areas, and the sky darkness according to the Bortle classification, mentioned occasionally in this book and explained below.

In February 2001, John Bortle published his now widely recognized sky darkness scale in Sky & Telescope magazine. It's a useful tool to help observers judge and describe the true darkness of the sky from a given site. It is based upon the actual visibility of given night-sky objects, and what might be seen through a standard medium-sized (12.5-in/32-cm) telescope. It is reproduced here with permission.

Fig. 1.10 The night sky's brightest star, Sirius, glitters above the Brecon Beacons Dark Sky Reserve in Wales. (Photo: Michael Sinclair)

First, a note on the term 'magnitude' as ascribed to a star.

The concept of a star's magnitude dates from classical times. The Greek astronomer Hipparchus, assuming the brightest stars to be the biggest (Latin *magnitudo*, size, bigness), divided them in his star catalog (thought to have been compiled about c.140 BC) into orders of brightness or magnitudes. Three centuries later, Ptolemy declared in his Almagest (c.140 AD) that the brightest stars would be of the first magnitude, prominent but less 'important' ones of the second, and so on. The faintest stars visible were to be of the sixth magnitude. All this was rather subjective, and hardly lends itself to twenty-first century scientific accord and accuracy, so nowadays it is agreed among astronomers that a star of magnitude +6.0 will be one hundred times fainter than a star of magnitude +1.0. It follows from this that a star of, for example, magnitude +5.0 is 2.512 times brighter (2.512 being the fifth root of 100) than a star of magnitude +6.0, and so on up and down the scale. The plus sign is important, since the magnitude of the brightest star in the night sky, Sirius, becomes by this reckoning −1.4. The bright star Vega's magnitude is 0.0. Astronomers retain this confusing scheme. Please don't ask why.

Bortle's sky darkness scale has nine divisions:

Class 1: Excellent Dark Sky Site The zodiacal light (see Chap. 4), Gegenschein, and zodiacal band are all visible, the zodiacal light (Fig. 1.11) to a striking degree and the zodiacal band spanning the entire sky. Even with direct vision, the galaxy M33 is an obvious naked-eye object. The Scorpius and Sagittarius region of the Milky Way casts obvious diffuse shadows on the ground. To the unaided eye the limiting magnitude (faintest star seen) is 7.6 to 8.0 (with effort); the presence of Jupiter or Venus in the sky seems to degrade dark adaptation. Airglow (a very faint, naturally occurring glow most evident within about 15° of the horizon) is readily apparent. With a 32-cm (12½-inch) 'scope, stars to magnitude 17.5 can be detected with effort, while a 50-cm (20-inch) instrument used with moderate magnification will reach 19th magnitude. If you are observing on a grass-covered field bordered by trees, your telescope, companions, and vehicle are almost totally invisible. This is an observer's Nirvana!

Class 2: Typical Truly Dark Site Airglow may be weakly apparent along the horizon. M33 (Fig. 1.12) is rather easily seen with direct vision. The summer Milky Way is highly structured to the unaided eye, and its brightest parts look like veined marble when viewed with ordinary binoculars. The zodiacal light is still bright enough to cast weak shadows just before dawn and after dusk, and its color can be seen as distinctly yellowish when compared with the blue-white of the Milky Way. Any clouds in the sky are visible only as dark holes or voids in the starry background. You can see your telescope and surroundings only vaguely, except where they project against the sky. Many of the Messier globular clusters (Fig. 1.13) are distinct naked-eye objects. The limiting naked-eye magnitude is as faint as 7.1 to 7.5, while a 32-cm telescope reaches to magnitude 16 or 17.

Class 3: Rural Sky Some indication of light pollution is evident along the horizon. Clouds may appear faintly illuminated in the brightest parts of the sky near the horizon but are dark overhead. The Milky Way still appears complex, and globular clusters such as M4, M5, M15, and M22 are all distinct naked-eye objects. M33 is easy to see with averted vision. The zodiacal light is striking in spring and autumn (when it extends 60° above the horizon after dusk and before dawn), and its color is at least weakly indicated. Your telescope is vaguely apparent at a distance of 20 or 30 feet. The naked-eye limiting magnitude is 6.6 to 7.0, and a 32-cm reflector will reach to 16th magnitude.

Fig. 1.11 The zodiacal light. (Photo: Nick Hart)

Fig. 1.12 M33, a spiral galaxy 2.7 million light years away, in the constellation of Triangulum. (Photo: Nick Hart)

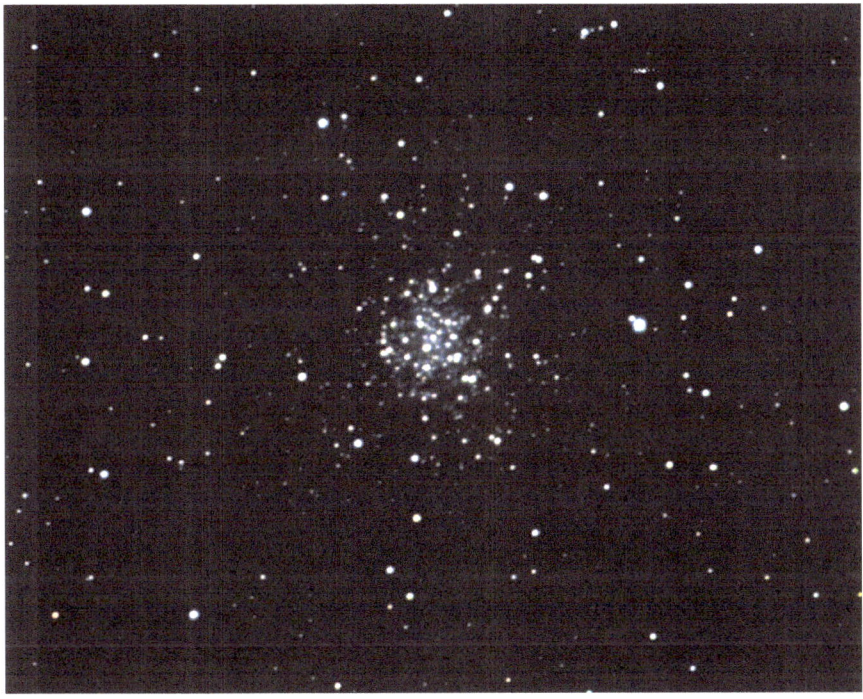

Fig. 1.13 Globular star cluster M56, 33,000 light years away. (Photo: Dave Finnigan)

Class 4: Rural/Suburban Transition Fairly obvious light-pollution domes are apparent over population centers in several directions. The zodiacal light is clearly evident but doesn't even extend halfway to the zenith at the beginning or end of twilight. The Milky Way well above the horizon is still impressive but lacks all but the most obvious structure. M33 is a difficult averted-vision object and is detectable only when at an altitude higher than 50°. Clouds in the direction of light-pollution sources are illuminated but only slightly so, and are still dark overhead. You can make out your telescope rather clearly at a distance. The maximum naked-eye limiting magnitude is 6.1 to 6.5, and a 32-cm reflector used with moderate magnification will reveal stars of magnitude 15.5.

Class 5: Suburban Sky Only hints of the zodiacal light are seen on the best spring and autumn nights. The Milky Way is very weak or invisible near the horizon and looks rather washed out overhead. Light sources are evident in most if not all directions. Over most or all of the sky, clouds are quite noticeably brighter than the sky itself. The naked-eye limit is around 5.6 to 6.0, and a 32-cm/12.5-inch reflector will reach about magnitude 14.5 to 15.

Class 6: Bright Suburban Sky No trace of the zodiacal light can be seen, even on the best nights. Any indications of the Milky Way are apparent only toward the zenith. The sky within 35° of the horizon glows grayish-white. Clouds anywhere in the sky appear fairly bright. You have no trouble seeing eyepieces and telescope accessories on an observing table. M33 is impossible to see without binoculars, and M31 (Fig. 1.14) is only modestly apparent to the unaided eye. The naked-eye limit is about 5.5, and a 32-cm telescope used at moderate powers will show stars at magnitude 14.0 to 14.5.

Class 7: Suburban/Urban Transition The entire sky background has a vague, grayish-white hue. Strong light sources are evident in all directions. The Milky Way is totally invisible or nearly so. M44 or M31 may be glimpsed with the unaided eye but are very indistinct. Clouds are brilliantly lit. Even in moderate-size telescopes, the brightest Messier objects are pale ghosts of their true selves. The naked-eye limiting magnitude is 5.0 if you really try, and a 32-cm reflector will barely reach 14th magnitude.

Class 8: City Sky The sky glows whitish-gray or orangish, and you can read newspaper headlines without difficulty. M31 and M44 (Fig. 1.15) may be barely glimpsed by an experienced observer on good nights, and only the bright Messier objects are detectable with a modest-size telescope. Some of the stars making up

Fig. 1.14 M31, the Andromeda Galaxy. (Photo: Nick Hart)

the familiar constellation patterns are difficult to see or are absent entirely. The naked eye can pick out stars down to magnitude 4.5 at best, if you know just where to look, and the stellar limit for a 32-cm reflector is little better than magnitude 13.

Class 9: Inner-City Sky The entire sky is brightly lit, even at the zenith. Many stars making up familiar constellation figures are invisible, and dim constellations such as Cancer and Pisces are not seen at all. Aside from perhaps the Pleiades, no Messier objects are visible to the unaided eye. The only celestial objects that really provide pleasing telescopic views are the Moon, the planets, and a few of the brightest star clusters (if you can find them). The naked-eye limiting magnitude is 4.0 or less.

Appendix 4 in this book shows the correlations between the Bortle scale and the readings obtained from a properly calibrated Unihedron SQM.

Fig. 1.15 Seen from the Isle of Mull, Scotland: bright Jupiter nears M44, the Beehive star cluster in Cancer. (Photo: Shaun Reynolds)

Now we will define the gold, silver and bronze tiers for IDSPs:

Gold Tier Near-natural conditions and limiting magnitude (faintest star seen) of 6.8; Bortle Class 3, SQM reading 21.89 – 21.69. Most people, when their eyes have become dark-adapted, can just make out magnitude 6 stars from a dark site. Remember, the higher the magnitude number, the fainter the star.

Silver Tier Exemplary night sky showing most of the features of a natural sky. Limiting magnitude 6.3; Bortle Class 3 or 4, SQM reading 21.69 – 21.25.

Bronze Tier Milky Way visible, limiting magnitude 5.8, and Bortle Class 4 or 5, SQM reading 21.25 – 19.50.

For countless years, on every clear night, human beings have been able to behold the spectacle of the starry heavens, spangled with thousands of stars and traversed by the Milky Way. Nature's grandest free show has spurred us to consider our place in the great scheme, has given rise to many themes and aspects of our cultures and religions, and has inspired both artistic achievement and scientific endeavour. Bob Mizon, Light Pollution: Responses and Remedies, Springer 2012.

Why are so many of us taken aback, yet at the same time awestricken, by the unaccustomed view with our own eyes of natural sky wonders that our distant ancestors knew? Why will we travel long distances to experience it again and again once we have discovered it? Hasn't technology nowadays supplied us with enough celestial images to satisfy our interest in the universe? Haven't the Hubble Space Telescope and other less well-known instruments, for example the Spitzer Space Telescope (www.spitzer.caltech.edu), given us, as they glide through Earth's cosmic environs, enough stunning photos to keep us sitting at screens indoors rather than traveling to star parties? Does direct sight of the glorious starry firmament awaken some deep, species-wide memory within us? Perhaps there is some half-formed realization when starwatching that we are indeed part of something so much vaster, and that we should be wary of accepting an immeasurably shrunken environment, and assuming that Earth is all that exists.

In October 2012, Superstorm Sandy ravaged twenty-four U. S. states, causing particularly severe damage in the New York City tri-state area. Power outages followed, and for the first time in many New Yorkers' lives, they were able to see the night sky.

America's most light-polluted skies are arguably those above the great urban agglomeration of the New York tri-state region. Only a handful of the brightest stars, the brighter planets and the Moon are usually visible, if there isn't a tall building in the way. Bortle classes 8 and 9 are just about everywhere.

When the lights went out, police began to receive calls from spooked citizens about strange lights in the sky. The callers were seeing stars for the first time. News items, echoing reports from a blacked-out Los Angeles in 1994, recorded alarmed residents describing mysterious, mottled linear 'clouds'—the Milky Way, their home galaxy, spanning the darkened sky. The contrast between the starscapes of rural Maine and those over the New York City area is discussed in Ian Cheney's award-winning 2010 documentary The City Dark, which is a minutely researched, clear-sighted, and really impressive film.

This author taught modern languages, and what little astronomy had found its way onto the UK national curriculum, for many years in a secondary-level school in Poole, in the county of Dorset, southern England. Poole is part of the biggest metropolis on the southern coast of England. Three seaside towns (Bournemouth, Poole and Christchurch) and their suburbs have gradually grown together since Victorian times to become a multi-centered sprawl housing more than half a million people. In summer, holidaymakers add many tens of thousands to the total. This is the largest urban area in the UK without the official title of city.

Fig. 1.16 Corfe Castle, Dorset: Comet Pan-STARRS shimmers below the Andromeda Galaxy. (Photo: Mark Gaston)

The school ran a summer camp each July for a year group of over a hundred students, fifteen and sixteen years old, and for a week they and several of their teachers lived in tents on the hilly chalk downland of the Purbeck area, to the south-west of the big town. Down the road, the stark, ancient ruins of Corfe Castle, established by William the Conqueror in the eleventh century and a victim of the English Civil War in the seventeenth (Fig. 1.16), stood silhouetted against the sunset glow in the evenings. From this site, distant town lights were screened by the high chalk ridge of Nine Barrow Down, and skies were, and still are, dark. As an amateur astronomer, I volunteered to be camp night-time supervisor (and unofficial astrophotographer) while teacher colleagues slept. Sleepless students would creep out of tents and use the telescopes and binoculars provided. They would marvel at the Milky Way (again, "What's that big cloud?") and exclaim in amazement about the richness of a night sky they had no hope of seeing from their urban homes. There are some bright lights in that unofficial city. One night, from a dark, wind-buffeted coastal car park, I read a newspaper by the light of floodlights at the Poole Channel Ferry port from a distance of 9 miles. Even kids not particularly sensitive or environmentally aware were mesmerized on those clear summer nights by what was, for some, their first encounter with their lost heritage—the vault of stars above.

Fig. 1.17 Members of the Wessex Astronomical Society prepare for a public observing session. (Photo: Robert Hatch)

The Wessex Astronomical Society is one of hundreds of groups of night-sky enthusiasts in the UK, and one of the biggest and most active. The society has a large observatory at Durlston Country Park, a dark coastal site above the cliffs near Swanage in Dorset. The facility was officially opened in 2009 by Member of Parliament Lembit Öpik, grandson of a famous Estonian astronomer, Ernst Öpik. There are public open evenings at the park several times a year, and when the Atlantic consents to hold back its clouds, the society shows visitors the stars through members' telescopes (Fig. 1.17) and the big 14-inch reflector beneath the observatory dome. A typical visitor count in 2009 would have been perhaps fifty or sixty people. Nowadays, an event such as a planetary conjunction, a new comet or a meteor stream can bring 150 or more observers, with, happily, a good percentage of children.

There is currently a definite upsurge of interest in astronomy. In many countries, astro club enthusiasts report increased public attendances at indoor and outdoor meetings. Telescope sales climb. Television features about the near and far cosmos no longer screen at off-peak times, but command prime slots. In April 2015 the Guardian newspaper reported on the increase in visitor numbers to the

Northumberland National Park/Kielder Forest area in northwestern England since it had achieved international dark sky status in 2013: "Many of the 1.5 million who visit Northumberland each year are now aware of its Dark Sky status. A car-hire firm in Newcastle says that a lot of people come to see the sky now, especially in autumn and winter when it's darkest. The hotels and guest houses are in business all year round."

California's Lassen Volcanic National Park has the dark skies of a high-altitude site, and is far enough away from large urban areas to attract star seekers. In 2012, in an effort to increase visitor numbers, the park instituted the annual Lassen Dark Sky Festival. This weekend event has seen, according to the park authority, "greatly increased visitor attendance and participation at dark sky programs. Park staff have found that people are more likely to travel long distances to the park for a multi-day event. The festival celebrates Lassen's dark skies and creates an opportunity for visitors to participate in astronomy-related activities during the day and connect to the park by hiking, sightseeing, and camping."

So, with the general increase of interest in things astronomical, the stars are shining brighter on the tourist agenda. Why do people so enjoy being beneath unfamiliar stars?

Half of our environment, the starry sky, is above the horizon. Undeniably an area of outstanding natural beauty and site of special scientific interest, this half of our environment has helped to mold the minds of humans. Our ancestors, on every clear night, could gaze in awe at the thousands of stars visible to the unaided eye. They incorporated the sight and the motions of the stars, Sun, Moon and planets into their cultures, their calendars, religions, art and emerging science. They invented fanciful stories to explain the ghostly white bridge across the sky that we now know to be our Milky Way Galaxy, the multi-billion-star spiral that carries the Solar System through the dark reaches of the cosmos. Could this line of ashen light above be a trail of cornmeal spilled by a hungry dog running from the grain store, as the Cherokee legend had it? Or perhaps the Pacific islanders knew better—it was a line of cloudy mud churned up by the constantly beating tail of a hunting celestial shark.

Sadly, the starry vault, our inspiration to so many things, has quietly and gradually disappeared over the last few decades. All across the developed world it is now largely unknown territory. The late Michael Crichton, author of Jurassic Park, wrote in his autobiography Travels[1]: "The natural world, our traditional source of direct insights, is rapidly disappearing. Modern city-dwellers cannot even see the stars at night. This humbling reminder of Man's place in the scheme of things, which human beings once saw every twenty-four hours, is denied them. It's no wonder that people lose their bearings, that they lose track of who they really are, and what their lives are really about."

I believe that, just as people enjoy walking among trees, which densely surrounded our vastly distant ancestors, so we savor, when we see it, the forgotten marvel of a star-strewn sky, an image held somewhere deep within our consciousness. The shock of the Milky Way, seen by a modern town-dweller for perhaps the first time, is that primeval wonder resurfacing.

---

[1] Pan/Macmillan, 1988, ISBN 0-330-30126-8.

A common reaction to an untainted night sky is what Robert Macfarlane called "a sudden flipped vertigo" and a feeling that you are very small, even insignificant, beneath the vault of stars. It may be unsettling to be told that the light from the faint smudge of M31, the Andromeda Galaxy, has taken 2.5 million light years to reach our eyes—and M31 is the Milky Way's nearest large neighbor in our Local Group of about fifty major and minor galaxies (Fig. 1.14 above). One remedy for this sensation of insignificance after a night at the telescope or simply relaxing beneath the stars on your sun-lounger is to make sure you have a microscope waiting at home to restore a sense of scale!

We shouldn't feel uneasy or overawed as we rediscover our heritage of spangled skies, but rather enjoy our re-acquaintance with the stars that made us and with the sight of the Milky Way, our 500-billion-star vehicle moving through a vast universe.

If you're fortunate enough to live in a rural fastness with a dome of thousands of stars overhead on cloudless nights, or you can visit a nearby site with such views, get out there at every opportunity. Don't miss a passing comet, meteors, the northern lights and other sky wonders.

This author lives on the northern edge of that three-town urban sprawl on the southern coast of England, yet my sky at night is acceptably starry, except on some nights when increased dust and haze in the atmosphere increase skyglow, as they scatter the upward emissions of town lights to the south. Why is my sky better than might be expected for an urban area? Because the local authority (Dorset County Council) has replaced all our local street lights with flat-glass types, which means that the lenses are parallel to the ground and allow the emitted light only downwards (Fig. 1.18). So from my back garden, on a moonless night, I can see, without telescope or binoculars, a few hundred stars, the Andromeda Galaxy and, faintly, the Milky Way when the Moon is absent.

Is there any way you can improve the view from your urban location? Chapter 8 will suggest ways.

If you're a beginner in astronomy, or just wondering what the night sky really looks like, try joining a local astronomy group, or just consulting its website. This may be a passport to a dark sky place. Many such groups have their own out-of-town observing sites and advertise their gatherings, inviting the public to come and share their enjoyment of the cosmos. Whether the location is a fully equipped observatory complex or just a level piece of accessible land, it's good to be with people who know how to find night-sky objects and enjoy showing them. Astronomy groups welcome beginners. Don't be afraid of being a small fish in a big pond—members like nothing better than explaining astro stuff!

You might like to go it alone (or, if of a nervous disposition, take friends with you) and seek out a rural spot where you can legally unfold a camping chair, set up a camera, telescope or mounted binoculars, and watch the skies. Local astronomers

Fig. 1.18 Full cut-off (FCO) lights, sending all emissions downwards. (Photo: Bob Mizon)

will know of such places, recommend routes, and give advice on the best sightlines and access. Find the biggest skies, away from tall trees and looming mountains. Reconnoiter the area beforehand in daylight to ensure that there are no thorn bushes, foot-traps, rabbit holes or even abandoned mine shafts! This is not a flippant remark—I was once walking a potential observing area in the Clee Hills of Shropshire, birthplace of the Industrial Revolution, when I came across the unfenced remains of an old small-scale coal mine, looking like a deep artillery-shell crater in a grassy field. See that very hole on shropshirehistory.com/mining/cleecf.htm.

What's the best time of year to observe? Some dates might offer more targets than others. Chapter 4 takes us on a tour of the sky through the seasons. Some phenomena are ephemeral and unpredictable. Consult astro websites for the possibility of aurorae or passing comets, usually faint binocular objects. If you'd like the reassurance of predictable events, meteors tracing their fiery paths in the sky are more frequent on certain dates. With a good star atlas and a red light to preserve night vision, you will soon learn your way around the heavens. A good webpage listing star atlases is www.astronomy.com/observing/get-to-know-the-night-sky/2014/04/choose-a-star-atlas-thats-right-for-you.

Planetarium programs (see Appendix 4 in this book) are readily available. For example, the free download Stellarium is an excellent guide to the motions and identities of sky objects. It has details such as the positions of the moons of Jupiter in real time, and images of deep-sky objects as seen through telescopes and by cameras. Zoom in on laptop or tablet and see close-ups.

Most programs have night modes, so they can be dimmed and in some cases reddened. Shield any light you produce, even if it's dim and red. Purists say that the use of these, even in night mode, can compromise your dark-adapted eyes. It's your choice as to whether to bring the technology or use charts (see Appendix 4 in this book) and a dim red lamp. If you use paper charts, laminate them or keep them in clear plastic envelopes to prevent them from absorbing moisture from the air. Keep books in plastic bags when not in use.

In Chap. 5, we'll look in more detail at what kit you will need for observing sessions.

Chapter 2

Invasion
of the Townies

Skyglow is denying people the opportunity to see the glory of a starry night, which is part of our environment and heritage. When people go to the countryside they are usually blown away by seeing the night sky for the first time.
  Graham Bryant, Commission for Dark Skies

The recent upsurge in people's interest in astronomy has meant nationwide publicity about the creation of dark sky places where stars are protected and stargazers are catered to. See, for example, the information and video on www.breconbeacons.org/stargazing. Increasing numbers of hopeful aurora watchers book northbound cruises; eclipse chasers travel halfway around the world to see the Sun disappear behind the Moon. People worldwide now participate online in projects such as Galaxy Zoo, planethunters.org and seti@home, using their computers to classify galaxies, and search for extrasolar planets and signs of extraterrestrial life.

The phenomenon of astro-tourism is the principal subject of this book. The wonders of the universe are in books and magazines and on screens. Ever-improving CGI techniques now make us all astronauts moving through space, exploring planetary surfaces and weaving through the debris of a collision between a space station and a rogue satellite. Little wonder that so many of us are inspired to see the cosmos for ourselves. Safaris beneath the stars are a growing reality. We want to relate to the universe directly, not just see the pictures.

© Springer International Publishing Switzerland 2016
B. Mizon, *Finding a Million-Star Hotel*, The Patrick Moore Practical Astronomy Series,
DOI 10.1007/978-3-319-33855-2_2

Any primary school teacher will tell you that there are two things in the early science curriculum that modern seven-year-olds can't get enough of: dinosaurs and outer space, both of them remote, magical, vast, tempting with the twin lures of the totally unattainable and the visually splendid. Such fascinations linger on in most of us.

Many well-publicized space events have happened in recent years, spurring our interest. After the Apollo lunar missions had us on the edges of our seats during the late '60s and early '70s, there was perhaps a 'lull after the storm': Not easy to follow that.

What managed to follow was the launch in 1990 of the Hubble Space Telescope, and its later performance in orbit. The HST captured the public imagination. The fact that its optics needed a 'fix' in 1993, with astronauts adding new optical components—'spectacles for the HST'—made the mission even more interesting. The vast Internet photo gallery of this wonder telescope expanded, and records huge numbers of 'hits.' One of the best known images in the world must be that of the iconic Pillars of Creation (1995), the 20-trillion-mile high dust towers of the Eagle Nebula, followed by a second, more detailed image dated 2015 (Fig. 2.1).

Fig. 2.1 The Pillars of Creation, in the Eagle Nebula. (Photo: NASA/ESA/Hubble Heritage Team (STScI/AURA)/J. Hester, P. Scowen, Arizona State University)

A memorable event, still talked about by non-astronomers, even though they often confuse its name with another of its famous celestial sisters, was Comet Hale-Bopp (not Halley-Bopp). This comet (Fig. 2.2) was bright and easily observable, even from cities, for over a year in the late 1990s, and was estimated to be about a thousand times brighter than Halley's Comet on its 1985-86 visit. Discovered by Alan Hale and Thomas Bopp in 1993, the brilliant comet had been curving towards us through space for thousands of years after passing through the deep-freeze of the remote Solar System. Its previous pass of Earth had been around 4,200 years ago. A possible record of this event exists in the darkness of an ancient Egyptian tomb, that of the pharaoh Pepi I (2332–2283 BC). In his pyramid at Saqqara is a text referring to a 'long-hair-star.'

The first landing on an asteroid (Eros) and the first tourist in space (American businessman Dennis Tito) made the news in 2001, and the Spirit and Opportunity Mars rovers were headliners in 2004. The landing of the Huygens probe on Titan in 2005 and the seemingly endless stream of Saturn images from its companion craft Cassini were viewed across the world. In 2015, we marveled at the amazing Rosetta mission, a hot topic in all media, dropping a probe on a comet called Churyumov-Gerasimenko (Fig. 2.3). Newscasters often referred to Comet Churyumov-Gerasimenko by its serial number, 67P, for obvious reasons.

The pinpoint engineering triumph that was the New Horizons Pluto mission, with the spacecraft sweeping in summer 2015 through the Plutonian system after nine years of hurtling past the planets, again focussed public interest in what lies out there (Fig. 2.4).

Eclipses make the news nowadays, and also make work for travel companies. On the morning of March 20, 2015, the Moon's shadow passed near the British Isles. Thousands left workplaces and schools to observe the Sun (through proper eclipse viewers, of course) as a thin crescent in the sky. More adventurous enthusiasts had boarded ships or flown to the Faroe Islands or Spitsbergen (Svalbard) to see the Sun totally hidden by the Moon.

A total lunar eclipse of September 28, 2015, was seen across most of the United States, and the total solar eclipse of August 21, 2017, will attract observers from all over the world to the track of the Moon's shadow as it crosses the continental United States from Oregon to South Carolina. A later chapter will explore this event in detail, and look forward to other solar eclipses up to and including the next major European eclipse in 2026, when the track of totality will pass across the Atlantic from Greenland, shaving the western coast of Iceland and making landfall in northern Spain.

So, with all this recent inspiration, we can understand why the starry sky, and the phenomena created by the Moon and our daystar the Sun, are pulling in more viewers.

Fig. 2.2 Comet Hale-Bopp graces the sky in 1997. (Photo: Peter Carson)

Fig. 2.3 Comet 67P/Churyumov-Gerasimenko. (Photo: ESA/Rosetta NAVCAM)

There are hundreds of websites on the subject of stargazing holidays. There are long lists of Dark Sky Communities, Parks and Reserves; Dark Sky Discovery sites (DSD is a UK network of organizations promoting astronomy and the environment). There are mentions of nighttime hikes and astronomy 'adventures'; festivals focusing on the mythology and science of the constellations and planets; 'sky-friendly' rural hotels and guest houses; campsites with telescopes and visiting astronomers; public-access observatories. All are inviting people to share their starry skies.

Local administrations' tourist offices and websites invite visitors to come stargazing, and publicize their dark-skies efforts. Here are some examples.

Fig. 2.4 Pluto from New Horizons. (Photo: NASA/Johns Hopkins University Applied Physics Laboratory/Southwest Research Institute)

From the Orange County (North Carolina) site, Recreations section:

Stargazing at Little River: Volunteers from Morehead Planetarium will bring high-powered telescopes and give a tour of the night sky. Bring a blanket, camp chairs, your own telescope or binoculars.

From Sustainable Lighting Strategy for Edinburgh (City Council of Edinburgh, capital of Scotland):

(Paragraph 1.2) Protecting darkness and dark skies is also important. The mission of the International Dark-Sky Association (IDA) is to preserve and protect the night-time environment and our heritage of dark skies through environmentally responsible outdoor lighting. Global initiatives like Earth Hour raise awareness of the effects of consumption. Adopting a lighting strategy can assist in coordinating and helping to reduce the impact of lighting, protecting the contrasts created by the atmospheric character of the city.

Even big cities can try to bring more stars to their nightscapes. One holiday company's website listed thirty-two rural places to stay, across the UK, which mention their good night skies, with some providing specialized facilities for astronomers. Given the pages and pages of websites that come up for "stargazing holidays" alone, there must be many hundreds of such places on what Bill Bryson calls our 'small island',[1] and far more across the greater expanse of the United States and elsewhere. Chapter 3 explores a selection of dark sky destinations in the UK and in the United States.

---

[1] Eleven U.S. states are larger than the United Kingdom. The UK is a little smaller than Oregon, and slightly larger than Wyoming. A fun way to compare the two countries is to drag the UK around the United States on www.sarmonster.net/UK.htm.

What astronomy experiences may be on offer at a dark sky destination? What range of facilities can you expect? Has your destination a lighting strategy to ensure good viewing? Is there anything else to do if the weather doesn't cooperate? If a dark observing site has no overnight accommodation, will there be some nearby? An excellent example of an informative dark sky brochure, addressing the above considerations, is to be found at www.beacons-npa.gov.uk. The Brecon Beacons National Park's *Dark Skies Guide* leaflet defines its status as Wales's first IDA Dark Sky Reserve, shows photos of dark skies, star trails and the aurora taken from the park, and has maps of the best locations within the reserve. Also, it features basic seasonal constellation charts and notes on the depredations of light pollution. Lastly, to add a little spice, it has a list of "Did You Know…" astro facts, i.e. did you know that we're moving through space at 67,000 miles an hour? Perfect publicity (Fig. 2.5). You don't have to search for long for what you need to know for your stargazing sortie.

Fig. 2.5 Dark sky literature issued by the Brecon Beacons National Park. (Photo: Bob Mizon)

Chapter 3

Seeking Out the Real Night Sky

Here the skies can be so dark that the Milky Way and even the planet Jupiter can cast shadows: a phenomenon many have never seen.

Kielder Observatory (UK) website

The lists of dark sky destinations that follow consist partly of places that this author has visited, and personal views are provided on their facilities. Others are included on the basis of descriptions and suggestions from fellow astronomers, astrophotographers and contacts in various parts of the UK and the United States, or from the owners or staff at the chosen locations. For each destination we have gathered as much information as we could on facilities, sky quality, and in some cases lighting strategies (some have no lights at all). After each entry relevant websites are cited, giving contact details and information about where to stay and what you might see and do at these places. The fact that there is a good venue doesn't necessarily guarantee clear weather and pristine views of the cosmos—just that it looks promising!

Save time and money by carrying out your own research before embarking on an astronomy trip. Going to northern Scotland to see the aurora in July? Don't bother—there's precious little darkness at that time of year there. Buying a large automated telescope to take to an astro-campsite? You could spend most of the time setting it up and aiming it rather than seeing sky targets. Is there one already provided, operated by an experienced person?

© Springer International Publishing Switzerland 2016
B. Mizon, *Finding a Million-Star Hotel*, The Patrick Moore Practical Astronomy Series, DOI 10.1007/978-3-319-33855-2_3

Urban safety issues mean that many open spaces in towns both in the United States and the UK are either closed at dusk or lit to such an extent that little can be seen in the skies from there. In less illuminated public spaces, starwatching with a group is recommended. Rural sites might offer more accessible locations.

Our current 24-hour society has created a situation where many people are unfamiliar with the dark and are uneasy with no lighting. The Home Office's Police.uk website echoes this fear of the dark, advising us to "avoid… walking across parks or commons" at night. There is little evidence that lighting levels and crime rates are connected, but unused areas such as car parks and green spaces are often lit throughout the night, so an urban open space is unlikely to have dark skies. Recent UK studies[1] have shown that, in areas where councils have switched off lighting at midnight, crime rates do not increase. Further evidence shows that they actually fall.

We'll concentrate on more rural areas where real darkness reigns. In a remote location you might consider starwatching with some human company. In sparsely inhabited areas phone signals may be non-existent. At least let someone know where you are going and when you expect to return. Consult weather forecasts. Contact local agencies to check on conditions of public access, road closures, and the possibility of the presence of not-too-friendly wild animals, especially in the United States, though the UK has its share of creatures you really don't want to disturb, or tread on, in the dark.

These lists of dark sky places are not, of course, exhaustive, though the suggested sites are among the very darkest. But many more will be found with a quick trawl of websites such as, for the UK:

www.britastro.org/dark-skies/bestukastrolocation.html
www.darkskydiscovery.org.uk
www.stargazing.holiday

and for the United States:

www.observingsites.com
www.darksky.org/component/content/article/36-ida/night-sky-conservation/
    91-darksky-finder-a-destinations
www.fodors.com/news/photos/10-best-stargazing-sites-in-the-us

National Parks, local and regional tourist offices and astronomical groups are also ready sources of information. Lists of UK and U.S. astronomy groups can be found respectively at:

fedastro.org.uk/fas/members/members-a-z
www.astroleague.org/al/general/society.html

---

[1] www.lshtm.ac.uk/newsevents/news/2015/reduced_street_lighting_does_not_increase_crime.html.

A detailed light pollution map of the UK, created by Frédéric Tapissier, president of Avex, a French organization for astronomy and information on light pollution, is available at www.avex-asso.org/dossiers/pl/uk/index.html

Philip's/CfDS Dark Skies Map of Britain and Ireland (ISBN 0-540-08612-6) is a folding map of sky visibility, with main roads and places shown.

The Need-Less Interactive Night Sky Simulator on www.need-less.org, created by CfDS member and graphic designer Dan Nixon, is a wonderful tool for searching out dark sky sites in the UK. Move the cursor around the country, zoom in on locations, and watch the Milky Way come and go as you raise and dim the lights: www.need-less.org.uk/images/ukatnightsim.swf.

An excellent light pollution map of the United States can be found at djlorenz. github.io/astronomy/lp2006, together with other highly detailed maps of light pollution on all continents.

A detailed interactive map from the U.S. National Park Service is available at www.nature.nps.gov/night/skymap.cfm.

Try also www.lightpollutionmap.info produced by the Earth Observation Group, NOAA's National Geophysical Data Center and Slovenian astronomer Jurij Stare.

Southern coastal counties of England have the clearest weather on average, with the best percentage of cloud-free or partly cloudy nights. On the whole, the northern half of the UK has just over half as much sunshine during the year as the southern. Prevailing winds are from the southwest, so areas further inland, and especially areas in the north, plus mountainous regions, have more cloud. The likelihood of cooler northeasterly winds increases in spring, and the occasional northerly or southerly flow can dramatically affect temperatures.

When large anticyclones move in from the Atlantic or Scandinavia at any time of year, long, clear spells are possible throughout the UK, with very cold nights in winter. Your best weather advisors for planning your starwatching are probably the following:

www.metoffice.gov.uk/public/weather/regional-forecast
www.metoffice.gov.uk/uk-cloud-cover-forecast
www.metoffice.gov.uk/climate

Scotland

Scotland has many areas where truly dark skies can still be found (Fig. 3.1). Its rugged and mountainous heart and extensive moorlands and forests ensure thinly populated regions, as do its islands. Cloud cover is more frequent than in England.

Fig. 3.1 Pristine night skies over southwest Scotland. (Photo: James Hilder)

Clearer weather is more likely, especially in spring and autumn, on the west coast of Scotland, the Hebrides and Orkney. The summer can often see clear skies, but remember that, the further north you go, the more likely it is that the night sky will never be completely dark in later spring and summer. Changeability is the watchword for Scottish weather. There is an old saying that some places can have "four seasons in a day." November to February, when nights are longest and darkest, has the most tempestuous weather, but if you're lucky, Scotland, mostly unpolluted by stray light, offers breathtakingly starry vistas.

The further north you are in the UK, the more likely it is that you will see the aurora, or northern lights. Astrophotographer Graeme Whipps has set up his camera in many Scottish locations (Fig. 3.2). The darkest skies he has encountered are in the northwest, for example in Benbecula, Gairloch, and Durness, a favorite with aurora hunters.

Twenty-five-year data from the British Astronomical Association's Aurora Section conclude that in the far north (Lerwick, Shetland), the annual frequency of northern lights displays is between 164 and 44, and in southern Scotland 90 and 6. Frequencies in the English Midlands are between 23 and 1, and in the south 10 and zero. Included in this book are a good number of photos of the northern lights, to try and dispel the myth that they never descend to lower latitudes, and that you must book an Alaskan or Norwegian-Fjords cruise or an ice-hotel holiday in Scandinavia to see them.

Graeme Whipps (c)

Fig. 3.2 Meteor and aurora from Pitcaple, near Aberdeen. (Photo: Graeme Whipps)

1. GALLOWAY FOREST PARK. In the Dumfries and Galloway region of southwestern Scotland is Galloway Forest Park. It is large, some 300 square miles (780 km^2) in size, with a very small population. By day, the park is a tranquil expanse of mainly forested tracts and rugged terrain (Fig. 3.3); by night, it is a primeval dark place where, with clear skies, countless stars cast their faint light upon the quiet, darkened scene below (Fig. 3.4). SQM readings of up to 24 (incredibly dark) have been recorded there.

   Galloway became the UK's first Dark Sky Park in 2009, after hard preparatory work by Forestry Commission Scotland and others. The northern part of the park is home to the Scottish Dark Sky Observatory. Opened in 2012, it has large telescopes, lecture facilities and an observing deck, and allows visitors to bring and use their own astronomical instruments. There is accommodation in nearby Dalmellington and Carsphairn.

   In the southern part of the park is the small town of Newton Stewart, in the welcoming Creebridge House Hotel, shortly before the area gained IDSP status. You can tour the area and enjoy its night skies from both the hotel car park and a private observatory deep within the forest park. Newton Stewart has several hotels and guest houses. The Galloway Forest Astronomy Society is based locally. Visitors to the area are very welcome to make contact, and the society can assist them with advice on observing the park's velvet-black night skies.

   www.gallowayforestpark.com
   www.scottishdarkskyobservatory.co.uk
   www.newtonstewart.org/accommodation.asp
   www.gallowayforest-astro.org.uk

2. KILMICHAEL GLEN, ARGYLL. Not all dark-sky destinations are extensive wildernesses. On a smaller scale than Galloway Forest Park is a very dark site in Kilmichael Glen, mid-Argyll: Comraich (Gaelic for 'sanctuary' and pronounced 'comree'). It is a private property on the west coast of Scotland, and

Fig. 3.3 Galloway Forest Park on a frosty morning. (Photo: Bob Mizon)

Fig. 3.4 The Cygnus region of the Milky Way over Galloway Forest Park. (Photo: James Hilder)

Fig. 3.5 The Singing Heart Cottage in Kilmichael Glen. (Photo: Sue Stubbs)

its owners Sue Stubbs and Dave Smith welcome stargazers in search of a better experience of the constellations. A 60 × 22 GreenKat spotting 'scope is available, and accommodation currently consists of a cottage (Singing Heart Cottage), a spacious static caravan, and plenty of room for camping (Fig. 3.5). Outdoor lighting is located underneath porches and is used only when necessary. The light control at this site earned it the British Astronomical Association's Good Lighting Award in 2015. Sue says that, although she can't guarantee that the weather will always cooperate, on clear nights it's *extremely* dark and very starry! Remember that this far north (56°N), from late spring until early autumn, skies will not get completely dark even around midnight. Kilmartin Glen, 2 miles from Comraich, has a notable concentration of prehistoric monuments, including stone circles and standing stones of astronomical significance.

www.thesingingheart.com (Comraic)
antiquity.ac.uk/projgall/scott324

3. THE ISLE OF COLL. The Isle of Coll became the world's fifth International Dark Sky Community in December 2013. Coll is in the Inner Hebrides and lies 6 miles (10 km) off the coast of Argyll in western Scotland. A population of about

200, its distance from the mainland and the island's exterior lighting management plan combine to ensure the darkness of its skies. Its bird sanctuary is one of three recommended observing locations.

The main township and ferry terminal is Arinagour. Coll's highest point at 340 feet/106 m above sea level is Ben Hogh, near a large ice-age erratic boulder, the Clach na Ban-righ, or Queen's Stone. Ben Hogh is another of the many places for stargazing on this elongated island. Organized activities include a stargazing weekend event called Coll and the Cosmos at Coll Bunkhouse. An Internet search will reveal others, as well as a video about the dark sky community, at the Bunkhouse's website listed below.

In the daytime visit Coll's archaeological sites: Iron Age forts, cairns and ancient dwellings, and the two standing stones of Tofronald, thought to have astronomical significance.

Photographer Ewan Miles' four-minute video celebrating Coll's stars can be found on vimeo.com/91336703. See also:

www.visitscotland.com/info/towns-villages/isle-of-coll-p1162401
collhotel.com/darkskies
collbunkhouse.com/dark-skies-coll-cosmos

4. THE ISLE OF LEWIS. In July 2015 the BBC reported plans to set up a whale-song listening station and astronomical observing site at Gallan Head, on the Isle of Lewis, larger than Coll and further to the north in the Hebrides. At the time of writing in early 2016, the Gallan Head Community Trust was pursuing this initiative, to redevelop disused military buildings on the island with a view to creating what Trust Chairman Martin Hayes described as "a multi-purpose observatory where visitors can study and enjoy the dark skies, the marine wildlife of Loch Roag, and the remarkable natural and historical environment." This forms part of the Cetus Project, in collaboration with the University of St. Andrews and Stornoway Astronomical Society. The project aims to attract visitors to observe wildlife and listen to whale-song (Cetus is the constellation of the Whale in the sky) by day, and enjoy the Milky Way and myriad stars at night. The nearby Gallan Head Hotel, described by the Netmums website as "arguably one of the nation's quirkiest eateries," is favorably reviewed on travel advice sites.

See the following websites. The Stornoway Astronomical Society's website has a very useful map of the best dark sky sites on Lewis and adjoining Harris.

www.isle-of-lewis.com
www.hebrides-news.com/star-observatory-plans-22915.html (Cetus Project)
www.stornowayastro.org/darksky.php

5. THE ISLE OF SKYE. Skye is the largest island in the Inner Hebrides and is connected to the mainland by a toll-free road bridge. Skye's mountainous core is dominated by the bare, rock-strewn peaks of the Cuillins (Fig. 3.6). The island flora is largely moorland heather, and Skye is known for its wildlife, for example red deer and golden eagles. On clear nights, skies can be pristine, and the

Fig. 3.6 A fireball illuminates the clouds over the Cuillin Mountains on Skye, October 2013. (Photo: Marcus Macadam)

lack of skyglow reveals countless stars. The Milky Way is visible in great detail for most of the year, though in spring it is low to the northern horizon.

There are nine dark sky sites on Skye.[2] Promotional organizations Destination Skye, Lochalsh and Visit Waternish have identified and publicized these on websites and with leaflets, and Forestry Commission Scotland (FCS) has made marker posts for the sites. Destination Skye and Lochalsh's Margaret Arscott recommend the Camas na Sgianadain site, level and easily accessible, with ample parking. Accommodation and facilities are available in nearby Broadford. Margaret and husband Gerry hold informal observing sessions on the Clan Donald estate. The Astronomer Royal for Scotland, John Brown, once attended, but unfortunately he brought the clouds with him! The weather can change very rapidly in this part of the UK, and visitors are advised to make themselves aware of local forecasts.

www.visit-waternish.co.uk/dark-skies.php
www.destination-skye-lochalsh.com
maps.forestry.gov.uk
www.metoffice.gov.uk/public/weather/mountain-forecasts/west-highlands

---

[2] Waternish, Trumpan car park, Stein Jetty car park, Knockbreck Primary School car park, Kylerhea Forestry Commission Scotland car park, Kinloch Forest Forestry Commission car park (about a mile from the road along a FCS track, Camus na Sgianadin (FCS), and Clan Donald Steadings front and back lawns.

6. LOCH MORE, Caithness. In the far north of mainland Scotland, there are many remote places where the stars can be seen in their full glory. Astrophotographer Gordon Mackie says that "Caithness and adjoining Sutherland afford countless dark spots remote from artificial lighting. With Caithness being relatively flat, high points, especially beside the coastline, offer wonderful 360-degree hemispheric views of the night sky." The Dark Sky Discovery Site at Castlehill Heritage Center, Castletown, just east of the small port of Thurso, the northernmost town on the British mainland, is "wonderfully dark (~21.5 on the SQM), giving splendid views of the Milky Way and even the zodiacal light."

   A prime site in Caithness is Loch More, through which the River Thurso flows on its way to, not surprisingly, Thurso. By the loch, Gordon's SQM has recorded 22.1, corresponding to Class 1 on the Bortle scale—what John Bortle called "an observer's Nirvana." If you stay in Thurso, which has several hotels and guest houses, drive down the B874 through Halkirk, then on via Bridge Street and a minor road southwards to Westerdale with its picturesque water mill. Continue for about 6 miles (10 km) on a single-track road (two-way traffic with occasional spots where you can pass, typical of this area), southwest towards the loch. You are unlikely to meet much traffic on this part of the journey. The terrain is such that daytime driving, though, might be advisable.

   To shorten the trip you might consider staying at the Ulbster Arms Hotel in Halkirk. On arrival at the loch in daylight, you will see little sign of the castle that once overlooked Loch More. Park near the loch and weir (barrier to divert water), by the old, unoccupied loch-keeper's cottage (Fig. 3.7), said to have once briefly housed Rudolf Hess after his mysterious flight to Britain in 1941. There is plenty of space here to set up a telescope or camera when darkness comes, or just deploy your recliner and raise your binoculars.

   Remember, midsummer observing is out of the question at this latitude, and, given the remoteness of Loch More, it's particularly recommended that you let someone know where you are going and your expected time of return.

   www.visitscotland.com/info/towns-villages/thurso-p243991
   www.caithnesian.co.uk/westerdale-mill
   www.ulbsterarmshotel.co.uk

7. THE ISLAND OF HOY. The island of Hoy, accessible by ferry (see Visit Scotland website below), is the second largest island in the Orkney archipelago. It is a place of dramatic coastlines, including the towering 449-foot (137-m) sea-stack of the Old Man of Hoy. This 55-square-mile (143 km²) island has a good number of hotels, hostels and guest houses. Roads are mostly confined to the eastern side of Hoy, with one cross-island route to Rackwick. Basing yourself at any of the guest accommodations listed on the website, find these roads and choose a safe place to park beside them. Investigate with a lamp first in case of sharp rocks or other hazards, then savor, and photograph if you feel adventurous, the star-strewn sky.

Fig. 3.7 The cottage at Loch More beneath the star clouds and dust lanes of the Milky Way. (Photo: Gordon Mackie)

Orkney in general is an example of a remote and intrinsically dark area where, because of the lack of regulation of lighting in the UK, there are localized light pollution problems. Looking east from Hoy onto Scapa Flow, you will see stray light in the distance from oil industry and shipping installations and from the towns of Kirkwall and Stromness. The marine energy sector adds its share, as do, surprisingly, fish farms, which may use subsurface lighting of fish cages. The Commission for Dark Skies' local officers in the islands report that, since the 1970s, when Orkney was a stargazers' paradise, light pollution has gradually been allowed to make its unwelcome mark.

www.visitscotland.com/see-do/island-hopping/orkney
www.hoyorkney.com/visiting/accommodation

8. SHETLAND. Shetland is nearer the North Pole than any other part of the British Isles. The island of Unst, at its northern tip, is at nearly 61°N, only 5° from the Arctic Circle. Anchorage, Alaska, is at almost the same latitude. Darkness never fully descends here in the summer. From mid-May to mid-July the "sim-

mer dim" (summer twilight) prevails, whereas in mid-winter there can be as little as 6 hours of daylight. Away from the main town of Lerwick there is little light pollution, but be warned that cloudy nights are far more common than clear nights in Shetland, and sea-fogs (*haar*) are common from May to September. However, as the Shetland Astronomical Society states, "when the clouds are gone the night sky can be utterly spectacular, and there's always a chance of seeing the northern lights, known locally as the Merrie Dancers." Through a typical winter, a keen observer checking the skies on clear nights might see several low-level auroral displays and possibly one or two more spectacular events (Fig. 3.8). Bear in mind that aurorae are hard to predict. There are websites showing the probability of displays, such as www.spaceweatherlive.com/en/auroral-activity/auroral-oval.

Fig. 3.8 A meteor decorates the aurora over Shetland. (Photo: David Gifford)

Visitors may also reap rewards from Shetland's daytime wonders (terrain, culture, wildlife)—but if they get the stars at night, what a bonus.

www.shetland.org/things/explore-nature/northern-lights
www.shetland.org/plan/accommodation

9. LOCH OF SKENE, ABERDEENSHIRE. In East Central Scotland, to the west of Aberdeen and near the village of Kirkton of Skene, is the Loch of Skene, a large lowland body of water artificially enlarged by a dam at its outflow. More than a century ago, Loch of Skene water drove the mill wheel at Garlogie, to the south of the loch. The mill is now an industrial museum. Not far to the north, near Pitcaple just off the A96 road from Aberdeen, is the intriguing ninth-century Maiden Stone, also known as the Drumdurno Stone after a nearby farm. It is a 3-m high Pictish standing stone bearing curious religious and mythological designs. Both these locations have been recommended as dark sky sites, and Graeme Whipps' photo of an incipient aurora above the Maiden Stone (Fig. 3.9) will show why. The Loch of Skene car park offers level space to set up a telescope or a chair or lounger for naked-eye stargazing, and there is a small parking area near the Maiden Stone with a path that leads to the monument. The stone may be covered in midwinter to prevent frost damage. Nearby Inverurie has several hotels, and there are many more in Aberdeen to the east.

Fig. 3.9 The aurora provides a beautiful backdrop to the Maiden Stone, near Inverurie in Aberdeenshire. (Photo: Graeme Whipps)

en.wikipedia.org/wiki/Loch_of_Skene
www.ancient-scotland.co.uk/site/111 (Maiden Stone)
www.walkhighlands.co.uk/aberdeenshire/hotels.shtml

10. GLEN NEVIS, HIGHLANDS. In the Lochaber district of the Scottish Highlands, and below Ben Nevis, the United Kingdom's loftiest mountain at 4409 feet (1344 m) runs the narrow valley of Glen Nevis, accessible from nearby Fort William. The glen's spectacular Steall Falls, Scotland's second highest waterfall, with a single drop of 390 feet (120 m), cascades into a scenic gorge. In 2009, Glen Nevis was nominated as the world's first Dark Sky Discovery Site, at the same time as Newbattle Abbey in Dalkeith, near Edinburgh. On clear nights, at a reasonable distance from Fort William, the glen has magnificent starscapes. A road runs southeast from the town for a few miles along the lower end of the glen, giving way to a track as the terrain becomes steeper. Advice on safe travel through the area may be had from the Fort William Information Center, or from the Glen Nevis Visitor Center. Helpful rangers are a good source of information concerning periodic stargazing events. Worthwhile observing sites include the visitor center car park, the Lower Falls further along the glen, and for the more adventurous, the forestry path towards the summit of Dun Deardail.

www.darkskydiscovery.org.uk/dsdsites.html
www.visitscotland.com/info/services/fort-william-information-centre-p333001
glen.nevis@highland.gov.uk

11. PENTLAND HILLS REGIONAL PARK. This 35-square-mile (90 km²) area of high pastureland, dark hills and planted woodlands was designated a Regional Park in 1984. It sits at the northern end of the range of the Pentland Hills, south of the city of Edinburgh in eastern Scotland. The Pentlands are popular with hill walkers and the more robust breed of cyclists, offering long views and dark night skies. Scattered reservoirs supply Edinburgh with its water. Astronomers who have visited this and other dark sites in eastern Scotland recommend several dark observing places within the park. Two of them are near the small town of Balerno: the Harperrig reservoir to the west of the town and, to the south, the Harlaw reservoir, near the Harlaw House Visitor Center with its car park. Directions can be found at stargazerslounge.com/topic/18954-sun-2nd-dec-harperrig-reservoir-lothian.

The Harlaw House Visitor Center closes mid-afternoon, and so offers no facilities for night-time visitors, but the car park can be used to set up telescopes or merely sit back and look up. The hills to the south raise the southern horizon somewhat. To lower that horizon and escape the lights of Edinburgh more completely, try setting up along the A702 road to the south of the hills.

Tourist accommodation for this area is mostly towards Edinburgh, and the VisitWestLothian websites have details of hotels and guest houses.

visitwestlothian.co.uk/see-do/walks/hard/pentland-hills-regional-park/
visitwestlothian.co.uk/see-do/walks/hard/harperrig-reservoir-east-
    cairns-hill/
www.pentlandhills.org/pentlandhills/info/3/discover/30/visitor_centres/2

Northern Ireland

The whole island of Ireland, especially in the west, has magnificent sites for seekers of glorious daytime scenery and striking night-time starscapes. Stevie Beasant, secretary of the Northern Ireland Amateur Astronomy Society (NIAAS), imagines a north-south line between Derry/Londonderry and Cork (Republic of Ireland): "In essence, many places to the right (east) of that line are not of the highest quality for stargazing, but the majority of the area to the left (west) is excellent."

The Philip's/CfDS Dark Skies map of Britain and Ireland and the online UK maps mentioned earlier show Ireland's starriest areas, including the dramatic darkness of the Kerry International Dark Sky Reserve in the far southwest. In Northern Ireland, away from urban centers such as Derry/Londonderry and Belfast, the starriest skies are to be found in Antrim, near the northeastern coast, and in a broad belt across the northwest, center and southeast of the province. Three places are listed where amateur astronomers have reported stunning night skies.

Northern Ireland's hilly terrain and its proximity to the Atlantic make it generally cloudier than most of England, but clear nights are more frequent here than in many parts of northern England and Scotland, especially from April through September.

1. BEAGHMORE STONE CIRCLES, COUNTY TYRONE. To the west of Lough Neagh and Cookstown in Northern Ireland are the Beaghmore Stone Circles, where starwatching can be a rewarding experience on clear nights. There is little pollution from distant artificial lights.

    The seven intriguing stone circles at Beaghmore were unearthed by peat-cutters in the late 1930s, and partially excavated from 1945 until 1949. This is Northern Ireland's most extensive arrangement of ancient stones. The 5000-year-old Bronze Age site is not far from Cookstown, along the A505 road and then northwards up a lane for about 3 miles. Explore the imposing alignments and concentrations of stones by day, and, if you are lucky enough to have a cloud-free night, stay and watch the universe revealed.

    Astrophotographer Martin McKenna's beautiful photo (Fig. 3.10) shows a magnificent auroral display captured there in February 2014. In McKenna's own words: "I walked carefully into the huge area that is Beaghmore Stone Circles. The place was pitch-black, and even though I had a head-torch I had to move with great caution, because beyond the short beam range of the light was total darkness, and ancient rocks everywhere. The sky was incredible, by far the best

Fig. 3.10 A colorful auroral display beyond the Beaghmore stones. (Photo: Martin McKenna)

I had seen that year, sparkling with countless stars. Orion glowed behind me near a single tree on a hill… The auroral colors were unreal. We saw striking red beams piercing through the green arc, and in the north there would be constant fresh outbreaks of beams near Ursa Major, with the most incredible orange color I had seen in years."

The director of Northern Ireland's excellent Armagh Observatory, Mark Bailey, knows the impressive skies over Beaghmore: "It is a wonderful site to visit at any time, whether by night or day. On a dark, clear night, the unusual stone alignments, circles and rows, which all appear to have been laid out with some underlying astronomical purpose in mind, provide a wonderfully evocative foreground to a glittering starscape that stretches from one horizon to the other." SQM measurements here on moonless nights can be as good as 21.2, a dark rural night sky on the Bortle scale.

About half an hour away is County Tyrone's main town, Omagh, with, perhaps unexpectedly, the Ulster American Folk Park and an annual bluegrass music festival. The Silverbirch Hotel is one of several hotels in town, and has been favorably reviewed.

www.discovernorthernireland.com/Beaghmore-Stone-Circles-Cookstown-P2949
www.nightskyhunter.com/BeaghmoreStoneCirclesAuroraFeb27th2014.html
www.omagh.gov.uk
www.silverbirchhotel.com

2. KILLYLANE RESERVOIR, COUNTY ANTRIM. One of the NIAAS' observing places is not far to the east of Ballymena. Just south of the A36 Ballymena-Larne road, with ample parking, is the easily accessible Killylane reservoir, on open moorland and popular with anglers. The Milky Way and the constellations are well seen from here, and the NIAAS calls the site "a beautiful observing location." Local astronomical groups meet here. The weather can be change-able and often cloudy, but skies during clear spells are well worth waiting for. Impressive photos of Killylane and its stars by Martin McKenna and others can be seen at www.nightskyhunter.com/PerseidMeteorWatch2006.html.

   More reports of Killylane and photos taken there are at eaas.proboards.com/thread/1438?page=1, and the NIAAS website is a good source of information about astronomical activities in the County. Hotels and campsites in the area are detailed on the websites below

www.eaas.co.uk
www.discovernorthernireland.com/Antrim-Visitor-Information-Centre-
   Antrim-P6885heycamping.co.uk/Ballyclare/Killylane_Reservoir

Fig. 3.11 The Elephant Rock faces the northern lights over the Antrim coast. (Photo: Martin McKenna)

3. NORTH ANTRIM COAST. Not far eastwards from the 40,000 ancient hexagonal basaltic blocks of the Giant's Causeway are White Park Bay and nearby Ballintoy Harbour. On clear, transparent nights along this "Causeway Coast" of Northern Ireland, these and many other excellent sites attract starwatchers from all over. There are superb sky views across the ocean, with stars down to the horizon. Look northwest, towards the open Atlantic, and northeast, where Rathlin Island, a major reserve for seabirds and seals, sits like a steppingstone in the narrow waters of the Sea of Moyle. Rathlin is easily visible from the Antrim coast by day, and the Mull of Kintyre in Scotland is just 11 miles (18 km) away. At night, Rathlin's lighthouses blink below the stars. Figure 3.11 is a wonderful image of the northern lights from the coast between Ballintoy and White Park Bay, with auroral activity behind Elephant Rock (on the left). In local legend the rock is a woolly mammoth petrified by fallout from an ancient volcanic eruption. From certain angles in daylight and silhouetted against the dusk it can look eerily like a mammoth. For advice on local accommodation ask the information centers in Bushmills, just south of the Giant's Causeway, and in Ballycastle, from where the Rathlin Island Ferry departs.

www.nationaltrust.org.uk/white-park-bay
www.discovernorthernireland.com/Ballycastle-Visitor-Information-Centre
www.discovernorthernireland.com/Bushmills-Visitor-Information-Centre

The Isle of Man

1. Mann, to use a local name, lies in the middle of the northern reaches of the Irish Sea. A straight line from Great Britain's southwestern extremity, Land's End, to its northeastern tip near John O'Groats passes through the Isle of Man. The island is 32 miles (52 km) from north to south and, at its widest, 14 miles (22 km) wide. It has a surface area of 220 square miles (570 km^2). Officially, Mann does not form part of the United Kingdom, and it is not in the European Union. It is a British Crown possession with independent administration. May is the least cloudy month of the year, and January the cloudiest.

   The enthusiastic members of the Isle of Man Astronomical Society promote dark skies (Fig. 3.12) and welcome visitors to their dark sky observatory on the first Thursday of each month. They organize occasional open to the public events, and a very popular evening astronomy meeting, held about six times a year, called 'Pie in the Sky.' Based on Snaefell, Mann's only mountain at 2034 feet (620 m), the event is accessed by electric railway to the summit, and participants have a meal followed traditionally by apple pie, hence the name of the event. The site www.iomastronomy.org/DSO/dso.html lists the best observing locations on Mann, and there are numerous hotels and guest houses on the island.

www.iomastronomy.org
www.iomastronomy.org/observatory
www.visitisleofman.com

Fig. 3.12 The Milky Way over the Isle of Man. (Photo: Kevin Deakes)

Wales

The Welsh Assembly, Wales's parliament, has for many years shown interest in keeping the country's skies dark. Assembly members have attended meetings organized by astronomical groups and dark sky areas, and have promoted star-quality lighting, for example on the A55 Expressway road along the North Wales coast, an initiative for which the assembly received the British Astronomical Association's Good Lighting Award. Wales's principal cities, Cardiff, Swansea and Newport, are on its south coast. To the north of these, the mountainous principality, with its multitude of forests and lakes, has an axis of protected dark sky zones covering around 1500 square miles (18 % of its land surface), including the Snowdonia and Brecon Beacons National Parks Dark Sky Reserves and the Elan Valley Estate Dark Sky Park.

   Wales has more rain than England and Scotland, but fewer rainy days than Northern Ireland. It is often cloudy, the clearest skies occurring from April to July. May is the sunniest month.

1. BRECON BEACONS NATIONAL PARK. In 2013, the Brecon Beacons National Park became Wales's first area, and only the fifth in the world, to be granted the status of an IDA International Dark Sky Reserve. The dark sky team at the park headquarters put in tremendous efforts to ensure good lighting practices in their area. In 2015, Brecon Beacons' Funding Development Manager Ruth Coulthard accepted the CfDS Award of Appreciation for all the hard work that went into the dark sky bid. According to the UK Ordnance Survey, "you'd be hard pressed to find a community which goes to such great lengths to ensure light pollution is reduced. The hard work is obviously paying off – on a clear night you can see just about everything from anywhere".

   Two of many recommended stargazing locations within the National Park are:

   The Usk Reservoir. This site has a car park and, in the words of Ruth Coulthard, is a "beautiful place to have a family picnic as well as an ideal place to enjoy outstanding dark skies" (Fig. 3.13). This large, level area allows easy setting-up of telescopes and is easily reached by road from nearby Trecastle.

Fig. 3.13 Looking south from the Usk Reservoir in the Brecon Beacons National Park. (Photo: Martin Griffiths)

The National Park Visitor Center (Mountain Center). The easily accessible Visitor Center is signposted in the village of Libanus, on the main A470 road, and is accessible day and night. The limiting magnitude (faintest star seen with the unaided eye) is 6.37, an outstanding value.

On the park's northwestern edge is Llanerchindda Farm Guest House, near Cynghordy. Single astronomers, families or larger groups observe from here, in an environment geared to astronomy, using their own telescopes or a 6-inch telescope for hire. The West of London Astronomical Society visits and recommends this site.

www.breconbeacons.org/about-brecon-beacons-darksky-reserve
www.breconbeacons.org/stargazing (I recommend the video)
www.breconbeacons.org/stargazing-learn
www.cambrianway.com/index.cfm?id=148 (Llanerchindda Farm)

2. ELAN VALLEY ESTATE DARK SKY PARK. In 2015, the Elan Valley Astronomy group, in conjunction with the IDA, Dŵr Cymru (Welsh Water) and the Elan Valley Trust, proudly announced the creation of Wales' first International Dark-Sky Park. It is the world's first privately-owned IDA Dark-Sky Park.

The Elan valley, in the very heart of Wales, nestles among the ancient Cambrian Mountains and consists of a spectacular 72-square-mile (186 km^2) landscape of lakes and dams, with wooded mountain slopes and high-altitude vistas.

The Elan website describes the "exceptional starry skies and natural nocturnal habitat, where light pollution is mitigated and natural darkness is valuable as an important educational, cultural, scenic, and natural resource."

This author stayed in the area while supporting its dark sky bid, and explored its glorious night skies with local astronomers Les and Kris Fry, prime movers in the process of securing the Elan Valley estate's silver-tier International Dark Sky status. I attended Elan's dark sky launch in October 2015, and Sorcha Lewis, the star-friendly ranger who organized for us a rather spooky tour of the inside of one of the massive dams that hold back the valley's waters (Fig. 3.14), emphasized the team effort by a great number of individuals that such initiatives involve.

The Elan website has details of accommodation in the valley and in the nearby town of Rhayader. The Penralley Guest House, in the town, is a friendly and comfortable place, as is the Elan Valley Hotel further up the road towards the Visitor Center.

Cambrian Safaris provides evening tours of the area and Elan Valley Astronomy organizes starwatching sessions with them.

www.facebook.com/EVastronomy
www.elanvalley.org.uk
www.penralleyhouse.com
www.cambriansafaris.co.uk

Fig. 3.14 From a dam in the Elan Valley, a sky saturated with stars. (Photo: Sorcha Lewis)

3. PEMBROKESHIRE. To the west of the Welsh conurbations and jutting out towards Ireland, 63 miles (101 km) away by ferry, Pembrokeshire (see earlier) offers dark skies. There are plenty of campsites and hotels within, or at an easy distance from the dark sky zones to the west and on the coastal areas of the county. Its tourism website tells us: "One of the best areas in the country to see the night sky is Pembrokeshire, and THE best place in Pembrokeshire to view the Milky Way is the designated dark sky location at the National Trust car park at Broad Haven South. When the sun goes down and the night sky darkens the whole universe reveals itself above. All you need to do to experience it is step outside." Broad Haven is on St. Bride's Bay, at the western end of the Pembrokeshire peninsula, and the websites below list various places to stay.

www.visitpembrokeshire.com/explore-pembrokeshire/gazing-at-the-stars
www.visitpembrokeshire.com/holiday-accommodation
www.visitpembrokeshire.com/explore-pembrokeshire/beaches/broad-haven-south/
www.walesdirectory.co.uk/St_Brides_Bay.html

4. SNOWDONIA NATIONAL PARK. Snowdonia National Park in northwestern Wales covers an area of 823 square miles (2130 km²) and is Wales' biggest National Park. The area takes its name from Snowdon, the highest mountain in Wales at 3560 feet (1085 m). Interestingly, the National Park has a 'hole' in the middle, a derestricted area, around the slate-quarrying center of Blaenau Ffestiniog, allowing it to pursue the development of light industries following the decline of the

Fig. 3.15 A wintry night at Llyn Idwal, in the Ogwen Valley, Snowdonia. (Photo: Nigel Ball, www.nigelaball.com)

slate industry. Snowdonia's spectacular daytime mountain scenery is matched on clear nights by stunning sights in its night skies (Fig. 3.15). Spend time at the intriguing Welsh Slate Museum near the northwestern edge of the park at Gilfach Ddu near Llanberis; then at dusk head for the stars over the northern tracts of the park, for example, the impressive Llanberis Pass. In December 2015, the park announced its International Dark Sky Reserve status within the IDA scheme, joining the Brecon Beacons National Park and Elan Valley Estate.

The extensive area of the Snowdonia National Park means that there are countless starry places where the go-it-alone astronomer might set up a camera, binoculars or telescope, and many dark places to stay, for example, the unlit Trawsfynydd Holiday Village at Bronaber, south of Trawsfynydd Lake at the center of the park. Small hotels and guest houses are plentiful. From the comfortable and welcoming Pen-y-Bont Hotel, in the southern part of the park, you can enjoy the view of the ancient glacial Tal-y-Llyn Lake (Llyn Mwyngil). Over this lake looms the dark mass of Cader Idris, the Chair of Idris, the giant who, in legend, used the mountain as a seat on which to recline and admire myriad stars by night.

www.visitsnowdonia.info
www.llanberis.com
www.logcabinswales.co.uk
www.penybonthotel.co.uk

5. POWYS: LAKE VYRNWY. On much of the Welsh side of the border between
England and Wales is Powys, the largest county in Wales. Its mountainous nature
means that Powys has the lowest population density of all areas of Wales, and no
very big settlements. The largest is Newtown, with fewer than 20,000 people.
Powys generally has little light pollution, though cloud cover is frequent. East of
the Snowdonia National Park is tranquil Lake Vyrnwy (Llyn Efyrnwy) National
Nature Reserve, known for the variety of its birds of prey. Occasional organized
stargazing events take place there. The Royal Society for the Protection of Birds
manages the reserve and has a Solar System scale model along its Rhiwargor
Falls walk. Local hotels and guest houses, for example, the Lake Vyrnwy Hotel
& Spa, Llanwddyn, at the southern end of the lake, advertise their pollution-free
starscapes. Astronomers familiar with the area record "inky dark" skies, with
SQM readings of up to 22 (Bortle's 'Nirvana').

> www.lake-vyrnwy.com
> www.rspb.org.uk/discoverandenjoynature/seenature/reserves/guide/l/
> lakevyrnwy

England

England's temperate maritime climate is typified by relatively warm summers and
often mild winters thanks to the warming effect of the Gulf Stream. Prevailing
southwesterly winds bring frequent cloudy spells, distributed throughout the year.
February and March are the driest months and October to January the wettest.

The northwest is England's wettest region. Southern coastal counties are the
sunniest areas, and northern, western and mountainous areas are the cloudiest.

Sometimes, anticyclones settle over the United Kingdom and can persist for
weeks. They bring clear skies and fewer clouds, colder nights in winter and hotter
days in summer.

For regional light pollution maps of England, find the *Night Blight* maps produced by the Campaign to Protect Rural England (CPRE) at www.cpre.org.uk/
resources/countryside/dark-skies.

We'll start with northern England.

1. NORTHUMBERLAND NATIONAL PARK/KIELDER FOREST. In December 2013 the
Northumberland International Dark Sky Park came into being. This park, in the
far north of England, is Europe's largest area of protected night sky, embracing an
area of 572 square miles (1483 km²) and including the Northumberland National
Park and Kielder Forest, with its world-class observatory (Fig. 3.16). Lead
Astronomer Gary Fildes, prime mover in the creation of the Kielder Observatory,
works with other full-time staff and many volunteers to share Kielder's skies with
visitors, and plans even more extensive facilities in the future.

The website www.visitkielder.com/visit/kielder-observatory gives information on Kielder's extensive public stargazing program. The Dark Sky Park has
IDA gold tier status, and its claim to be "the best place in England for people to

Fig. 3.16 Star trails above the Kielder Observatory. (Photo: Gary Fildes)

go to enjoy the heavens" is hard to contest on a clear night when the sky is transparent, with stars so numerous it's hard to pick out familiar constellation figures.

In the southern part of the park, the intriguingly named Once Brewed National Park Visitor Center is another focus for information and observing. The center is located within the Hadrian's Wall World Heritage Site, and runs a public stargazing program of indoor presentations and outdoor on-site astronomy events using its own 16-inch telescope.

Several of the UK's Dark Sky Discovery Sites[3] are within the park, but as its website points out, almost any open area within its boundaries is good for observing. The website visitnorthumberland.com/Northumberland-Dark-Sky-Park offers "developed sites, like Walltown and Cawfields, and remote sites like Bulby's Wood in the Breamish Valley and Elf Kirk in Kielder Forest. The Stargazing Pavilion at Stonehaugh was built through a unique collaboration between the local community with Newcastle University School of Architecture and Kielder Water and Forest Park Development Trust's Art and Architecture program." Northumberland Tourism awards "Dark Sky Stay and Gaze" designations to local hotels and inns where dark skies are present or nearby.

At the edge of Northumberland National Park, in Wark, the Battlesteads Hotel is probably the only hotel in Britain to boast an on-site observatory with Dark Sky Discovery Site status. It has modern telescopes of various sizes and

---

[3] www.darkskydiscovery.org.uk/dark-sky-discovery-sites/map.

25×100 mounted binoculars specifically for guests with wheelchairs, and there are courses for all levels of expertise. Owner Richard Slade said: "We are delighted to offer our guests the chance to discover our beautiful skies. Astrotourism is an emerging and growing leisure market. With such a world-class environment for studying the stars on our doorstep it made sense to embrace this and add to the many reasons to visit Battlesteads and the county of Northumberland."

www.kielderobservatory.org
www.visitnorthumberland.com/darkskies
www.battlesteads.com/observatory
www.visitkielder.com/stay (local hotels, campsites and other accommodation)

2. NORTH PENNINES AREA OF OUTSTANDING NATURAL BEAUTY (AONB). The English poet W. H. Auden called the North Pennines his "great, good place," and the visitor may well agree with him. In the center of the far north of England, this high, open and timeless terrain of cascading streams and rivers and peat and heather moors bears reminders of an industrial past. The North Pennines AONB lies where the Pennine chain of mountains, the "backbone of England," gives way to the Kielder Forest and Northumberland National Park. The area is also a European and Global Geopark.

   Ninebanks Youth Hostel, near Hexham in Northumberland, is, according to staff members, "a great place to gaze skywards." Open all year, the four-star stone building invites bookings from individuals, couples, families and groups for beds, rooms or the whole hostel, and provides "good real food." The hostel entertains a range of people interested in what dark skies have to offer, and there are regular stargazing events as well as the chance to contemplate the universe, weather permitting, just by yourself in unaccustomed darkness. Not far away is another recognized Northumberland Dark Sky Discovery site, Allendale Golf Course.

   Drive beneath the stars along the B6277 road (Fig. 3.17), a 33-mile (53-km) route across the AONB between Alston and Barnard Castle, via Middleton in Teesdale. A friend who lives in Barnard Castle and travels the B6277 regularly at night describes "incomparable views" of the heavens from points along this route.

   www.northpennines.org.uk
   www.ninebanks.org.uk
   allendale-golf.com/?page_id=94
   www.sabre-roads.org.uk/wiki/index.php?title=B6277

3. THE LAKE DISTRICT. The Cumbrian Mountains and several large associated lakes radiating from their center are northwestern England's most popular tourist destination. Here in the Lake District National Park, England's largest, are the biggest and deepest lakes in England, respectively Windermere (Fig. 3.18) and Wastwater, and its highest mountain, Scafell Pike. Important towns are Keswick and Bowness-on-Windermere. The English National Trust's camping website describes Lake District skies as "stunning." In the park are many places in its

Fig. 3.17 The B6277 and a distant Moon. (Photo: David Williams)

Fig. 3.18 Looking down on Windermere. (Photo: Catherine Shreeve)

northern and eastern parts along roads (A591, A592) where the stars can be enjoyed just by pulling over at roadside parking areas. Almost anywhere away from settlements there are low levels of light pollution. If you're not wedded to hotels (and there are plenty in the area), you might try Wasdale campsite, near Scafell Pike, or one of many others far from towns on the park's website. Located in Ennerdale, England's most remote valley, is Low Gillerthwaite Field Center, a Dark Sky Discovery site hosting stargazing events throughout the year. The center, 2 miles from the public road, is accessed by a forest track beside Ennerdale Water. Its seventeenth-century farm buildings have self-catering accommodation for up to 40 people. On clear nights this wonderfully dark and tranquil area has night skies that most of the UK's population can only dream of. Friends of the Lake District, an organization promoting the protection of natural scenery in the area, has published a stargazing guide (see below). Weather in the Lake District can change rapidly: the National Park runs a forecast service called Weatherline: www.lakedistrictweatherline.co.uk. See these other websites, too.

www.lakedistrict.gov.uk/visiting/wheretostay
www.nationaltrust.org.uk/article-1355791742335
www.lgfc.org.uk/Visitors-DarkSky.htm
fld.org.uk/see-the-stars.html

4. NORTH YORKSHIRE: DALES AND MOORS. The southwestern flank of England's largest traditional county, Yorkshire, has more than its fair share of urban areas such as Sheffield, Leeds and Bradford. Further north and east, past York and Harrogate, population numbers thin and skies are darker. The National Parks of the North Yorkshire Dales and Moors have many starry sites. Don't go too far northwards, or the light domes of Darlington, Stockton and Middlesbrough will appear from over the county border in the River Tees Valley. Staying at the Grove Hotel in Leyburn, with access to both the Dales and Moors to its east and west respectively, you can find wonderful sky vistas in many places in the area, and watch Orion rise majestically in autumn from a dark, star-filled eastern horizon on the A6108 road to the south of the town. Local astronomer Richard Darn describes the Yorkshire Dales as "a stargazer's paradise: the Milky Way arches majestically across the sky on clear autumn evenings, and star clusters sparkle like diamonds on black velvet."

There are many places along the routes between the Yorkshire Dales National Park and the North Pennines AONB (see above) where darkness reigns, and you can simply stop and see a sky full of stars. Dark Sky Discovery 'Milky Way' class sites nominated by the Yorkshire Dales National Park Authority include Hawes, Malham and Buckden. In the words of park authority's Dark Skies officer Tracey Lambert: "We've known for a long time just how special the skies are above the Yorkshire Dales National Park, but it's nice to have this recognized. It's amazing what you can see – the Perseid meteors are spectacular and earlier this year it was possible to see the Northern Lights."

The Scarborough and Ryedale Astronomical Society (SARAS) will be a good source of information for stargazing on the moors. SARAS organizes public stargazing events. Many sites along the Cleveland Way National Trail, skirting the moors, are remote yet accessible by road. Pause on your way east-wards to the Dales to visit Hardraw Force, a 100-foot waterfall on the Hardraw Beck in Wensleydale. Interestingly, it is approached through a pub from the road. The Pennine Way long-distance footpath passes close by. You could base yourself in old market towns like Pickering, in the Dales, or Helmsley, on the moors, near the imposing remains of the abbeys of Byland and Rievaulx.

www.yorkshire.com
www.yorkshiredales.org.uk
www.northyorkmoors.org.uk
www.grove-guesthouse.com
www.scarborough-ryedale-as.org.uk
www.nationaltrail.co.uk/cleveland-way

5. YORKSHIRE WOLDS. In east and north Yorkshire are the low hills of the Yorkshire Wolds. The largest town is Driffield. In the churchyard at Rudston, 10 miles northeast of Driffield, is Britain's tallest ancient standing stone, a 40-ton megalith erected around 1600 BC. Another stone that will certainly interest the astro-tourist is the Wold Cottage meteorite. In December 1795, at about 3 p.m., a large chon-dritic meteorite, the second heaviest ever recorded in England, was observed fall-ing on land close to The Wold Cottage (Fig. 3.19) near the village of Wold Newton in East Riding. Major Edward Topham, owner of the cottage, investigated the fall and took statements from eyewitnesses. The bulk of the meteorite is now in the London Natural History Museum, and an obelisk stands on the site of the fall, bearing the inscription: "Here on this spot, December 13, 1795, fell from the atmosphere an extraordinary stone. In breadth twenty-eight inches, in length thirty-six inches, and whose weight was fifty-six pounds. This column in memory of it was erected by Edward Topham, 1799."

The Wold Cottage is now an award-winning bed-and-breakfast hotel, with additional self-catering accommodations and camping facilities. Guests have access to the meteorite obelisk, and owners Katrina and Derek Gray welcome astro-tourists at all times of the year. The hotel is 7 miles from the coastal bay of Filey, and the steep chalk cliffs of Flamborough Head and Bempton, famed for their seabirds, notably puffins. There are still plenty of tracts with low light pollution in the Wolds area, and there have been good reports of the night sky in and around Tophill Low Nature Reserve, between Beverley and Driffield, south of Wold Newton. The reserve closes at 6 p.m.

www.yorkshire.com/discover
www.driffield.co.uk
www.woldcottage.co.uk
tophilllow.blogspot.co.uk/p/visiting.html

We now turn our attention to the English Midlands.

Much of the western half of the Midlands is urbanized. In most of this area, dark skies are ruled out by the presence of England's second largest city, Birmingham,

Fig. 3.19 The Wold Cottage. (Photo courtesy Katrina Gray)

and nearby big towns such as Wolverhampton and Coventry. The large county of Shropshire is in the far west of the Midlands and shares a border with Wales. Shropshire, with few big urban areas, is the fourteenth largest county in England but ranks 45th in terms of population. Away from the light pollution of the Birmingham area, and of Telford and Shrewsbury, there are quiet rural areas in Shropshire along the border with Wales and to the south near Worcestershire. Here there are many places where stars abound.

The East Midlands has many large towns such as Leicester, Nottingham, Derby and Northampton. Its darkest areas are in its eastern half between Lincoln and the coast, around the Lincolnshire Wolds AONB.

If touring in the East Midlands, it's worth a detour to see fragments of Britain's most famous comparatively recent meteorite, which broke up in the atmosphere over Leicestershire on Christmas Eve 1965, scattering itself across Barwell and Earl Shilton, to the north of Hinckley and half-an-hour's drive westwards from Leicester. This chondritic meteorite, the largest known to have fallen in the UK in recent times, is estimated to have weighed around 46 kg (101 pounds) before fragmenting, and there are substantial pieces on display in Leicester at the New Walk Museum and the National Space Center, in Coventry at the Herbert Museum and in London at the Natural History Museum. For photos of some of the largest recovered fragments of the Barwell Meteorite, see piclib.nhm.ac.uk/results.asp?image=056606 and www.theherbert.org/collections/natural-history/rocks-and-minerals.

1. CARDING MILL VALLEY AND THE LONG MYND. The Shropshire Hills AONB, in southwestern Shropshire and extending into the hills of Wales to its west, is an extensive area of heather-covered uplands embracing 310 square miles (802 km^2). Church Stretton, an historic market town, lies at the heart of this area of the English-Welsh border, known as The Marches. The main A49 road and the Welsh Marches Railway Line run through the AONB, ensuring easy access. Camping and caravanning sites can be found on websites below. Those preferring hotel stays can find plenty in Church Stretton, not far from starry skies.

   Visitor facilities in Carding Mill Valley include dining, shops and good car parking. The valley is a very tranquil place. Not far away is the Long Mynd plateau, 7 miles (11 km) long and at maximum 3 miles (4.8 km) wide. This is a place of heath and moorland, with steep-sided valleys and gentler slopes. The rugged Stiperstones range extends to its west and the Stretton Hills and Wenlock Edge lie to the east. Much of this area is managed by the UK National Trust. Pathways both broad and narrow cross these areas, and from the top car park at Carding Mill Valley (a site mentioned by the Shropshire Astronomical Society as a good dark sky location), you can explore the vicinity and admire its stars. Further west, the darkest part of Shropshire is probably its western tip, either side of the B4368 road as it approaches the Welsh border along the valley of the River Clun.

   www.shropshirehillsaonb.co.uk
   www.churchstretton.co.uk
   www.shropshire-astro.com/main/home-page/shropshire-dark sky-reserve
   www.walkingbritain.co.uk/walks/walks/walk_b/3308/

2. THE LINCOLNSHIRE WOLDS AONB. The relatively flat landscapes of Lincolnshire are interrupted to the northeast of Lincoln by the more rolling terrain of the Lincolnshire Wolds AONB. This land inspired much of the work of the poet Tennyson, Queen Victoria's poet laureate, born in Somersby in 1809. Astronomers come to this area every autumn to attend the Horncastle Astronomy weekend, organized for the last 31 years by one of the UK's best known amateur astronomers, Paul Money. Horncastle is just to the south of the AONB, where night skies are, in the words of a member of a local astronomy group, "possibly the best in the county" (Fig. 3.20). Members of the East Lincolnshire Astronomy Club and their chairman Paul Cotton are happy to welcome visitors to their meetings at the Badger Farm campsite near the village of Asterby, halfway between Horncastle and Louth. The site is part of Lincolnshire Wildlife Trust's Red Hill nature reserve. Paul Cotton writes: "Here, the rural location gives splendid views of the Milky Way, with the Andromeda Galaxy being clearly visible to the naked eye, and M13, the Hercules globular cluster (see Fig. 4.21), visible with averted vision on nights of good seeing." Parts of the Lincolnshire Fens, further to the south, have reasonable skies at night. For his public viewing sessions Paul Money uses the Gibraltar Point National Nature Reserve Visitor Center near Skegness, Tattershall Castle, Chambers Farm Wood and, in the

Fig. 3.20 A green auroral curtain from Red Hill Nature Reserve in the Lincolnshire Wolds. (Photo: Paul Cotton)

north of the county, Far Ings Nature Reserve. Contact Paul at www.astrospace. co.uk, and see these other websites:
www.lincswolds.org.uk
www.lincsastro.co.uk
www.lincstrust.org.uk/gibraltar-point

We now turn our attention to eastern England.

1. NORTH NORFOLK (1): Kelling. The stars shine brightly over many places along the rural north coast of the large county of Norfolk, and its famously flat land-scape ensures big skies. The Kelling Heath Holiday Park, near Holt in the central part of the east-west coastal strip, has excellent night skies and hosts twice-yearly Equinox Star Parties (Fig. 3.21) in conjunction with the Loughton Astronomical Society. The autumn event, in September, is Europe's largest star-watchers' gathering. Hundreds of astronomy enthusiasts attend, and the site becomes a forest of tilting telescopes as night falls, the muted conversations of starwatchers joined occasionally by the sounds of a preserved steam railway that runs nearby. Lighting is strictly controlled. Day visitors and overnight campers are welcome, and there are two areas, with no-lights regulations, exclusively for astronomers at these times. If you'd rather observe in comparative isolation, there are many nearby beaches and roadside spots where the skies are equally tempting. If tents or mobile homes are not your style, try the Maltings Hotel, Weybourne, a couple of miles northwards along a narrow rural lane. Here there

Fig. 3.21 The skies of Kelling: the Moon and Mercury below the Pleiades. (Photo: Peter Carson)

is ample parking, a friendly welcome, traditional furniture, and starry skies from both the car park and the nearby Weybourne beach. One of the ales on offer close by at Weybourne's Ship Inn is called Moongazer.

starparty.las-astro.org.uk
www.kellingheath.co.uk
www.maltingshotel.co.uk

2. NORTH NORFOLK (2): Cley and Fieldview. Cley-next-the-Sea, also near Holt in North Norfolk, is another good place to enjoy sea and stars together. Nearby, the historic eighteenth-century windmill at Cley (sometimes locally pronounced "Cly") offers accommodation: nine unique rooms and an impressive circular sitting room with antique furniture. Upstairs, find extensive views over the marshes and the sea. Owner Simon says that "observation is possible from the walled garden and balcony around the Mill, unpolluted by lighting" (Fig. 3.22).

Fig. 3.22 Cley Mill beneath the stars. (Photo: Tim Doyle/DK Photography)

Near East Barsham in North Norfolk is Fieldview, another welcoming venue, run by Christine and Simon and offering self-catering holidays from Easter to the end of October. It sleeps up to eight people, and is 2 miles from Walsingham, well known for its religious shrines. There are telescopes for guests' use, and in the daytime the area is popular with bird-watchers, cyclists, hikers and naturalists.

In 2009–2010 the North Norfolk Astronomy Society used light meters and cameras to assess sky darkness across large parts of the North Norfolk area. The best reading obtained was 21.81 east of Bircham Newton, 12 miles north-east of the town of King's Lynn and possibly Norfolk's darkest site. The Breckland Astronomical Society's observatory at Great Ellingham, Norfolk, is an accredited Milky Way Class Dark Sky Discovery site.

www.cleywindmill.com
www.fieldview.net
www.nnas.org/o/darkskies/darkskies.php
www.brecklandastro.org.uk

3. Suffolk coast. Recommended by a colleague in the British Astronomical Association is unlit Walberswick on the Suffolk coast, within the Suffolk Coast and Heaths AONB. Astronomer Mike Gainsford states that "the area around Walberswick has a micro-climate, sometimes having clear skies when a few miles inland it's cloudy or raining." The nearby Suffolk Coast National Nature Reserve is occasionally used for organized star parties, and all along the coast

Fig. 3.23 The Milky Way, Walberswick. (Photo: Shaun Reynolds)

from south of Lowestoft to Aldeburgh and Orford there are places with good sky views, especially to the east where the Moon, stars and planets rise out of the North Sea as the land narrows to become the English Channel. The Milky Way is well seen from many coastal sites at the right time of year (Fig. 3.23). Stars become fewer in number as you descend the coast towards the obtrusive lights of Felixstowe, Suffolk's southern port and the UK's largest and busiest container destination. The Orwell Astronomical Society of Ipswich (OASI) gives advice on stargazing in the county. The Breckland Astronomical Society runs a star party during new Moon periods at Haw Wood Farm, near Saxmundham. The farm switches off its lights for the event.

www.thesuffolkcoast.co.uk
www.explorewalberswick.co.uk
www.oasi.org.uk

4. Dedham Vale. Dedham Vale, which harbors one of the UK's smaller AONBs, lies on the Essex-Suffolk border in eastern England. The River Stour, coursing from west to east through the vale, was depicted by the artist John Constable, in for example the most popular of all English paintings, *The Hay Wain*. The area is known as Constable Country. Tracts of ancient woodland punctuate landscapes of meadows and agricultural fields.

  Away from the A12, the major road east of the vale, skies are dark. The area around Flatford has good observing sites, and lanes will lead you to likely spots, though remember to obtain permission from any landowners whose premises you might want to cross. The Dedham Vale Society has produced guidance on lighting in this sensitive landscape and is working towards International Dark Sky Community status. As part of the society's early preparations for its bid for this status, members toured the area in February 2016 with a sky quality meter. Values of over 21 (dark rural sky) were recorded in the center of the area, though, as happens in many dark countryside areas in crowded England, light domes from distant towns taint the sky from the periphery. You can stay at the Crown Hotel at Stoke by Nayland, a modern and comfortable lodging which has rooms from which you can walk out through exterior doors onto lawns with good night sky views.

Sudbury is Dedham Vale's largest town. To its north is Lavenham, once the home of Jane Taylor who, in 1805, wrote the well-known children's song, *Twinkle, Twinkle, Little Star*. Could it have been inspired by perfect pre-Victorian skies over the Vale?

www.dedhamvalestourvalley.org
www.crowninn.net/the-hotel

5. THE DENGIE PENINSULA. "If you get as far away as possible from London and sky transparency is excellent, parts of Essex can surprise," says CfDS committee member Dan Nixon, who lives in the county. According to several local astronomers, the darkest place in Essex is the Dengie peninsula, in its southeastern part, between the two tidal rivers of the Crouch and the Blackwater. The Dengie National Nature Reserve is a popular bird-watching area. Essex people seem to pronounce Dengie either way, as *den-ghee* or *den-gee*.

Peter Carson, observing director of the Castle Point Astronomy Club (CPAC) in southeast Essex, has obtained SQM readings of 20.3 here, with magnitude 6 stars easily seen, and Dan Nixon has achieved 20.85. These are encouraging values for locations in the southeast of England. Much of the peninsula is accessible and offers good sky views, though the eastern tip is private land. Access to good observing spots there may be possible. To enquire about this, contact Carson through his Astromania website, and have a look at the Astronomy Club's website for more local information. There are a number of camp sites and hotels in and around nearby Burnham-on-Crouch.

Finchingfield, inland to the northwest and not far from Braintree, has sometimes been called "England's most beautiful village," though I'm sure residents of many other picture-postcard villages the length and breadth of the land might offer debate on this. The area around here has some admirable dark skies, and Nixon has recorded an SQM reading of 20.85 from nearby Stambourne. A base for exploring the area could be the Three Tuns Inn in Finchingfield; a search for 'Finchingfield accommodation' will reveal many hotels and pubs with guest rooms in the area.

www.astromania.co.uk
www.cpac.org.uk
threetunsfinchingfield.com

Southwest England is our next stop on this tour.

1. EXMOOR NATIONAL PARK. In 2008, Steve Owens, soon to become a stalwart of the organizing team of the International Year of Astronomy 2009, discussed with the Exmoor National Park's authorities the idea of protecting its night skies. In October 2011 (see Chap. 1), the park became Europe's first International Dark Sky Reserve within the IDA's Dark Sky Places scheme. After hard work by park staff, local astronomers, lighting professionals and the inhabitants of Exmoor, this largely unspoiled area followed Galloway Forest (Dark-Sky Park) and Sark (Dark-Sky Community) as a recognized sanctuary for the starry sky. Members of the CfDS joined the park authority team in 2010 and 2011 to take dark-sky measurements. Using Unihedron SQMs, readings of up to 22 (Bortle 1 dark!) were recorded at various locations.

There are many places in the inner core of Exmoor, which is crisscrossed by small roads (Fig. 3.24), where the casual stargazer can stop and contemplate some of the best night skies in England. The dark sky core zone measures 31 square miles (81 km^2), with minimal human habitation and many open-access areas. It has Bronze Age burial mounds, a Site of Special Scientific Interest at Dunkery Horner Wood National Nature Reserve, an abandoned medieval village at Hoccombe Combe, and areas used for astronomy events. In the buffer zones surrounding the core, the Exmoor National Park Authority works with

Fig. 3.24 Northern stars circle above Blagdon Cross, on Exmoor. (Photo: Adrian Cubitt)

Fig. 3.25 A Wimbleball Lake sunset. (Photo: Julia Amies-Green)

park residents and management of Somerset and Devon counties on dark-skies issues, with strong support from the national Campaign to Protect Rural England.

Star watches are occasionally held at the picturesque Wimbleball Lake (Fig. 3.25), and inquiries can be made through the park website. Wimbleball's campsite, says manager Alex Forster, is "ideal for overnight stargazing." West Withy Farm, near the lake, offers telescopes for on- or off-site use, and hosts astronomy experience nights. Dark Sky Telescope Hire works with Exmoor National Park Authority to help promote the reserve and has 6-inch Dobsonian telescopes (see Sect. 5.1) on loan from the three main national park centers in Dulverton, Dunster, and Lynmouth. The Dark Sky Telescope Hire website is a good source of information about Exmoor's night sky, accommodation and telescope availability

www.exmoor-nationalpark.gov.uk/visiting/things-to-do/stargazing
www.swlakestrust.org.uk/lakes-and-facilities/the-lakes/wimbleball-lake
www.exmoor-cottages.com
www.darkskytelescopehire.co.uk/exmoor-dark-sky-experience

Fig. 3.26 AstroAdventures: trees silhouetted against the aurora. (Photo: Zoe Ball and Dave Ward)

2. ASTROADVENTURES, NORTH DEVON. Along much of North Devon's border with Cornwall, and from nearby coasts, dark skies still reign. Light pollution maps of the area show an axis of darkness reaching down from the north towards the southwest peninsula's southern shores. In northwest Devon's Ruby Country and within easy reach of coastal resorts such as Bude and Bideford, Murray and Valerie Barber and Tony Gibbons run AstroAdventures, established in 2005. On a tranquil rural site near the lakes of the River Tamar, this is a venture that combines local holiday opportunities with the chance to observe the Milky Way in good skies, and examine detail in galaxies, nebulae and planets through large telescopes. Its absence of light pollution is of course a welcome feature for keen astro-tourists (Fig. 3.26), as are its unobstructed southern horizons. Both in winter and in summer, the south is the best direction to find celestial favorites: the Orion Nebula shimmers during the colder months, and in June, July and August far south constellations such as Sagittarius and Scorpius offer numerous clusters, nebulae and crowded Milky Way starfields. The AstroAdventures venue offers a large 20-inch (50-cm) Dobsonian telescope, a dedicated imaging observatory and a 7-inch (18-cm) telescope. Accommodation is comfortable, private and self-catering, with families in mind.

3. EDDINGTON LODGE ASTRONOMY. Based 5 miles northeast of Bude on Cornwall's northern coast, with minimal light pollution and the Milky Way easily visible, Eddington Lodge specializes in astronomy breaks (Fig. 3.27). Describing the sky quality there as excellent, owner Hannah says that their astro equipment is all on site and that all observing is done from the premises, with two observation pods and one main observatory with astrophotography equipment. To the north lies the Hartland Peninsula, and a search for "Hartland dark skies" will reveal the best stargazing spots there and suggest places where visitors can borrow telescopes and join organized astronomy events. The North Devon Astronomy Society meets monthly in nearby Barnstaple and may be consulted on observing opportunities in the area. See the websites below for more information:

www.eddington-lodge.co.uk/astronomy
www.northdevonastronomy.co.uk

Fig. 3.27 Eddington Lodge telescope. (Photo courtesy Eddington Lodge)

4. THE ROSELAND OBSERVATORY AND COURT FARM CAMPSITE/TOLCARN OBSERVATORY, CORNWALL. Brian Sheen's Roseland Observatory sits in a 4-acre field within the 50-pitch Court Farm campsite, west of St. Austell in Cornwall. Light levels are kept very low. In a *Sunday Times* survey it was once voted "one of the best 100 places to visit for starwatching in the whole of Europe." Light-meter readings give a naked-eye limiting magnitude of 6.0. Court Farm is set in 30 acres of tranquil pastureland, with 4 acres designated for caravans. Various telescopes (the largest a 16-inch Dobsonian) and binoculars are available both in the observatory and at the campsite, and there is an activity room with projection facility plus a meteorite and geological collection. During the season the observatory runs a regular open evening for visitors and also special sessions for groups and families by arrangement. The site has a weather station.

Tolcarn Observatory, also near St. Austell, is an accredited research facility with a special interest in planetary systems orbiting stars other than the Sun (exoplanets). It offers excellent facilities for observing, photography and live streaming. Director Grant Mackintosh is an experienced astronomer and astrophotographer, specializing in spectroscopy. Tolcarn has a number of instruments, including a 14-inch Meade LX 200, housed in two domes. "Glamping" and bed-and-breakfast are available at Tolcarn throughout the year. There are also stands for visitors' own telescopes.

For visits to both observatories, book through the websites.

www.roselandobservatory.com
www.courtfarmcornwall.co.uk/star-gazing.html
tolcarnobservatory.com

5. THE NORMAN LOCKYER OBSERVATORY (NLO) AND PLANETARIUM. One of the great personalities of Victorian/Edwardian astronomy was Sir Norman Lockyer, who discovered helium in the Sun and became an important pioneer in solar physics. The observatory that bears his name still stands in Devon, in the southwest of England, near the small seaside town of Sidmouth. An enthusiastic team of volunteers has developed the site in recent years, and it is now a prime location and focus for amateur astronomy activity in the region. The observatory holds public open evenings (just turn up 15 min before the published start time), and special group bookings for up to 60 people can be arranged.

Every year in early August the busy South-West Astronomy Fair draws crowds of astronomers, traders and exhibitors from all over the country to the NLO. Its modern planetarium, lecture theatre and display spaces are then busier than ever.

Chairman David Strange, whose photo of the aurora from the area (Fig. 4.26 in the next section) is evidence that the northern lights can indeed be seen from the south coast of the UK, reports that there are various local guest houses, one of which, the Longhouse, was Sir Norman's office and laboratory. Sidmouth, not far down the hill from the NLO, has plenty of hotels. Among local rural guest houses is Lower Pinn Farm. The front porch light goes out when the last guest is in, and, from behind the stables on the southern side of the premises, the summer Milky Way streams majestically down through Sagittarius and into the sea. Check out these websites:

www.normanlockyer.com/visitingus.html
www.visitsidmouth.co.uk
www.lowerpinnfarm.co.uk

6. THE ISLES OF SCILLY. Mike Glenny, who contributed some of his excellent astro-photography for this book, cautions astro-tourists not to forget the Isles of Scilly, Britain's most southwesterly outposts, defying the Atlantic 28 miles off Land's End at the tip of Cornwall. St. Mary's is the largest of the five inhabited islands (there are about 140 others) in the Scillonian archipelago. Mike says: "When I'm staying on St. Mary's, my favored observing site is on the south side of the Garrison. From this elevated wild spot, there is a clear view to the south over the sea and the low island of St. Agnes, while a hill behind blocks the light from nearby Hugh Town. Lighthouses and navigation markers in the distance don't detract from the views of the Milky Way, which on a clear night are stunning. As it's not normally necessary or appropriate to bring a car to the islands, I've never brought any serious astronomical equipment but have managed some wide-angle astrophotography with a fixed tripod." On St. Mary's is a hotel astrono-mers might like to consider:—the sixteenth-century Star Castle is laid out in the shape of a star.

Local astronomer Steve Sims recommends the following places on the main islands for dark sky observing: on St. Mary's, Garrison Field; on St. Agnes and St. Martin's, the cricket grounds; on Tresco, the playing fields; and on Bryher, by the community center. The low sea horizons are ideal for spotting Mercury, which never strays far from the rising or setting Sun. The rich southern starfields around Sagittarius and Scorpius, often lost to observers further north on the mainland, are better seen here.

Enjoy by day the summer flowers for which the Scillies are famous, and view their star-filled skies at night.

www.visitislesofscilly.com
www.islandspartnership.co.uk

Next stop, south-central England!

1. THE NORTH WESSEX DOWNS AREA OF OUTSTANDING NATURAL BEAUTY. The North Wessex Downs AONB, astride the rivers Thames and Kennet in the center of southern England, is the third largest of the 46 AONBs in the UK. Farmed for

the last 5000 years, the area is famed for its traces of historic and prehistoric peoples. Barrows, henges and white horses carved in chalk downlands mark the landscape alongside Roman roads and tracks more ancient still.

In October 2012 the AONB issued its position statement on energy. Its stated aims include initiatives "to conserve and use energy as efficiently as possible…, such as reducing external lighting and reducing highway lighting (to) benefit the AONB through protection of its dark night skies." This protection results in the area being an island of admirable starscapes away from the towns of the so-called M4 corridor along the London-South Wales motorway. At the western end of the AONB is the renowned megalithic complex of Avebury, now part of a World Heritage site, described by the seventeenth-century antiquarian John Aubrey as "exceeding in greatness the so-renowned Stonehenge as a cathedral doth a parish church." Debates continue as to the supposed astronomical alignments of some of the many large and small stones set in the earth around the village, but the sheer scale of the site, its easy accessibility and atmospheric feel—especially on misty autumn mornings or at dusk with a starry backdrop—will make a visit memorable.

Newbury Astronomical Society and Swindon Stargazers are among several organizations that observe within the AONB. Among recommended dark sites is Barbury Castle on the northern edge of the Marlborough Downs, from which many of the largest megaliths of Stonehenge originate. You might base your visit in Marlborough, an historic and pleasant market town to the west of the center of the AONB. It has plenty of accommodation among its half-timbered façades and old coaching inns.

www.northwessexdowns.org.uk
www.stone-circles.org.uk/stone/avebury.htm
newburyastro.org.uk
www.swindonstargazers.com
www.marlboroughwiltshire.co.uk

2. DURLSTON COUNTRY PARK. Durlston is on the Dorset coast near Swanage, in south-central England. It has a wide southerly view out across the English Channel. A winding path leads from the car park to the castle, following a timeline through 4.6 billion years of Earth's history. Durlston Castle overlooks the sea, and in the park there are various wildlife trails and associated activities. As mentioned earlier, it is the publicly accessible observing site of the Wessex Astronomical Society, with a learning center and a 14-inch telescope in a dedicated dome (Fig. 3.28). The society and Durlston's efficient team of park rangers host public open evenings several times a year. Members bring telescopes and large mounted binoculars, and groups of eager visitors visit the dome. Meetings are timed to coincide with sky events such as meteor displays, planetary conjunctions and (apparently) close approaches of planets to the Moon. Sometimes the absence of the Moon is preferred, so that the Milky Way and faint deep-sky objects are well seen. Outdoor parts of the site are freely accessible at night at other times.

Fig. 3.28 The dome at Durlston. (Photo: Arthur Lawrance)

Durlston has various wildlife trails and associated activities. After sunset in spring and summer the Milky Way appears as darkness falls, standing tall on the sea horizon, reflected in the water on calm nights. Durlston has wonderful starry skies, with little light intrusion from the Poole/Bournemouth area to the east. The nearest town to the south is Cherbourg in France, 83 miles away, faintly visible through telescopes when atmospheric conditions are right. Details of Durlston's public astronomy evenings are on the Wessex Astronomical Society's website listed below. Nearby Swanage, a coastal resort within easy walking and cycling distance, has large numbers of hotels and campsites.

www.visit-dorset.com
www.durlston.co.uk/visit-astronomy.aspx
www.wessex-astro.org.uk
www.virtual-swanage.co.uk

3. THE CRANBORNE CHASE AREA OF OUTSTANDING NATURAL BEAUTY. The Philip's/CfDS Dark Skies Map of Britain and Ireland shows the extent of light pollution in the UK using color coding. Big cities appear bright red, and pitch-dark areas are shown in gray. There are several in-between colors. Blue denotes Bortle Class 4 areas, where, although distant large towns may color the horizon, the Milky Way is still visible and the sky above is good and dark. The southern

counties of Dorset and Wiltshire have plenty of the 'blue' skies at night. The map shows the darkest place in south-central England to be the Cranborne Chase AONB, which straddles both counties and juts into Hampshire. It is a 380-square-mile (984 km^2) expanse of rolling downlands (open areas of chalk hills), copses of trees, ancient mounds capping burial sites of long-dead tribal chieftains, and chalk meadows of breathtaking summer wild flowers (Fig. 3.29). The AONB, in combination with the members of the Wessex County Astronomical Society, is in the process of measuring sky darkness and taking part in public starwatches and promotional seminars, with a view to establishing it as yet another of Britain's areas where the stars will have official protection. There are many places in the heart of the AONB where the casual observer can park off the road and see a strikingly beautiful night sky.

One particularly attractive dark site in the area is Knowlton Church and Earthworks, a few miles north of Wimborne, where the stark, weather-worn

Fig. 3.29 A quiet corner of an ancient landscape: Cranborne Chase. (Photo: Bob Mizon)

Fig. 3.30 The stars wheel above the ruins of Knowlton Church. (Photo: Nigel Ball)

ruins of a twelfth-century Norman church are surrounded by a large and much older Bronze-age henge. Nigel Ball's atmospheric photo (Fig. 3.30) captures something of the mysterious and historic feel of the place at night. There is 24-hour public access into the site from the road, through a small gate. Other good dark sky locations include Martin Down National Nature Reserve, which advertises its lack of light pollution, to the southwest of Salisbury, one of several such reserves, and Badbury Rings, an Iron Age hill fort near Wimborne. The National Trust manages Badbury Rings, and organizes occasional star watches there with the help of local astronomers. For information on these events, contact nearby Kingston Lacy House, a large and striking Italianate mansion and family home, open to visitors and famous for its paintings, carvings, gardens and the largest private collection of ancient Egyptian artifacts in the UK.

www.ccwwdaonb.org.uk/our-work/dark-night-skies
www.english-heritage.org.uk/visit/places/knowlton-church-and-earthworks
www.martin-pc.newforest.gov.uk/4527
www.visit-dorset.com/things-to-do/attractions/badbury-rings-p1244393
www.nationaltrust.org.uk/kingston-lacy

4. NEW FOREST NATIONAL PARK. In 1999, Sandy Balls Holiday Village, in the darker northern half of the New Forest National Park, earned the British Astronomical Association's Good Lighting Award for upgrading and shielding all its lighting to ensure a better view of the night sky for visitors.

This large and intriguingly named tourist destination is near Fordingbridge in Hampshire, southern England. The New Forest is one of the UK's most recently created national forests, receiving its status in 2005. It is a beautiful area of rolling heathland, wooded tracts and attractive small towns and villages, and is famed for its pony population (Fig. 3.31), who might be encountered during an observation period. When the ponies realize they can't eat your equipment, they move on peacefully. The southern part of the National Park is sandwiched between two large urban areas, Southampton to its east and Bournemouth/Poole/Christchurch to its west. The proximity of these two, and the big Fawley oil refinery on the southeastern border of the forest, mean that the southern half of the park has in many parts somewhat degraded skies (Fig. 3.32); but its northern half, away from the busy A31 main road that runs east-west across it, offers some very acceptable starscapes, and the local Fordingbridge Astronomers use sites there for dark sky observing sessions, to which visitors are welcome. Inquire through their websites for dates and dark locations away from villages and the busier roads. The National Park has many hotels and campsites throughout its 219 square miles (566 km^2). Towns within its borders include Lyndhurst, Brockenhurst and Burley.

www.newforestnpa.gov.uk/info/20001/visiting

Fig. 3.31 The New Forest by day. (Photo: K. D. Johnson)

Fig. 3.32 Wasted light from a golf driving range, tainting the night sky for miles around. (Photo: Bob Mizon)

www.sandyballs.co.uk
www.fordingbridgeastro.org.uk

5. THE ISLE OF WIGHT. Concerned people on this diamond-shaped island off the center of England's south coast have been fighting back robustly against light pollution. The Vectis Astronomical Society (Vectis Insula was the Roman name for the island) launched its Isle of Wight Dark-Sky Initiative in May 2013.

Following the example of Exmoor, Galloway Forest Park and many other sites in the UK, the society is pursuing IDA Dark-Sky status for a protected zone in the southwest of the island. The VAS collaborates closely with the island's AONB, systematically measuring the darkness of the night sky and promoting good lighting. Its observatory is open to all every Thursday evening, and it holds monthly public meetings in Newport.

The Isle of Wight branch of the Campaign to Protect Rural England (CPRE) has also turned its hand to spreading the dark sky message, promoting star counts and presenting annual good lighting awards to local organizations with carefully controlled lamps.

You can take the ferry to Yarmouth in the northwest of the Isle of Wight, where the British Astronomical Association's Good Lighting Award was presented to the Sentry Mead Hotel at Totland Bay in West Wight (Fig. 3.33). This was done at the suggestion of a fellow astronomer, who had stayed there and was delighted (or de-lighted?) that the owners switch off lights outside so guests can admire the night sky. The western and southwestern parts of the island have good dark skies with southern vistas of the Milky Way.

At Fort Victoria Country Park, near Yarmouth, is the Island Planetarium, incorporated into a Victorian red-brick fort. It was the first Dark Sky Discovery site on the Isle of Wight. Paul England and his team welcome visitors to their

Fig. 3.33 The Sentry Mead Hotel's Good Lighting Award. (Photo courtesy Sentry Mead Hotel)

360-degree star experience, and hold stargazing evenings and weekends in late summer through to late spring. Among their displays is one devoted to the microscopist, astronomer and inventor Robert Hooke, born in 1635 on the island. Hooke, who has been called 'England's Leonardo,' made some of the earliest detailed and accurate sketches of Saturn's rings and the craters of the Moon.

wightastronomy.org
www.darkwightskies.com
adsabs.harvard.edu/full/2010JBAA..120..337 M (Sentry Mead Hotel award)
www.thestarattraction.co.uk (Planetarium)

6. SOUTH DOWNS NATIONAL PARK. Towns like Portsmouth, Brighton and Chichester light the skies along the central strip of southern England's coastline, but to the north of them, the South Downs National Park remains a place of starry nights, since the geography of the area puts ridges, or downs, between the light sources and the inner parts of the park. Its dark-skies website tells us that "anyone who has driven the wooded lanes around Bepton and Petworth on a moonless night knows how dark it gets (Fig. 3.34).

On the likes of Butser Hill and Blackdown, amateur astronomers regularly set up their telescopes to explore night skies considered some of the best in the country." Dark Skies Ranger Dan Oakley states that "the South Downs NP has been measuring sky quality across the landscape and has discovered large areas of intrinsic darkness, surprisingly close to the large population that surrounds the park. With views of the Milky Way and the Andromeda Galaxy, the dark skies of the downs offer visitors a chance to engage with nature on a galactic scale. Capitalizing on the largest street lighting upgrade in the UK, we hope to show that, even in the brightest parts of the country, dark skies exist and are waiting to be enjoyed." In 2013 the park authority announced its intention to make the area an IDA International Dark-Sky Reserve, and has been working with residents and organizations within its borders to ensure star-quality lighting practices. Local residents can protect the area's stars by signing the park's dark-skies pledge. Hotels and camping facilities abound in this area of seaside towns and nature trails. Nearby Chichester has the wonderful South Downs Planetarium, for would-be stargazers to enjoy the heavens from a comfortable (ex-Boeing 747!) chair on cloudy days. Check out these websites for more information:

www.southdowns.gov.uk/national-park-authority/south-downs-centre/
www.visit-hampshire.co.uk/things-to-do/countryside/south-downs-national-park
placestovisitsussex.co.uk/areas/chichester-district/south-downs-national-park
www.southdowns.org.uk (Planetarium)

Fig. 3.34 The bright star Capella and the Pleiades bracket St. Hubert's Church at Idsworth on the South Downs. (Photo: Dan Oakley)

Fig. 3.35 The Hampshire Astronomical Group's Clanfield Observatory complex. (Photo: Graham Bryant)

7. CLANFIELD OBSERVATORIES, HAMPSHIRE. In the hills above the village of Clanfield, which is near the main A3 road from Portsmouth to London, is the Portsmouth Water Company's reservoir compound. What draws large numbers of people to the site on clear nights is not the reservoir but the state-of-the-art observatory complex (Fig. 3.35), with a learning center, five operational domes and one of the biggest public-access telescopes in southern England, a 24-inch reflector in an impressive housing. The observatory is owned and operated by the Hampshire Astronomical Group, one of the biggest and most active astronomy organizations in southern England; its chairman Graham Bryant reports that Clanfield's night skies have improved since the re-lighting of Hampshire with better-directed lamp types in 2013–2014. Many people who have never seen the Milky Way or a good starry sky come up from Portsmouth and surrounding urban areas, and are impressed by views of our Galaxy and the constellations above the observatories. Public observing evenings are so well subscribed that prior booking is necessary through the group's website (below). The observatory may be visited only by prior arrangement.

A virtual tour of the site is at www.hantsastro.org.uk/tour/index.php. Book your place at one of the group's popular observing evenings at www.hantsastro.org.uk/openevenings/index.php

And now for southeast England…

1. ROMNEY MARSH. Probably the best night skies in the southeast of England are to be found around the East Sussex/Kent border near the coast, in a circular area pivoted around the ancient Cinque Port town of Rye, which offers a good selection of hotels. The darkness that once cloaked the activities of smugglers along this coast nowadays draws in starwatchers. Big skies above low horizons, relatively low cloud frequency and limited light pollution make Romney Marsh a good observing venue, with the Milky Way visible on moonless nights. SQM readings of 21.3 (Bortle Scale 4) have been taken near Old Romney. The Wadhurst Astronomical Society, which has carried out local dark sky measurements, holds occasional events in dark sky areas and may be able to advise further on good locations for observing.

   During winter months public viewing evenings at the Romney Marsh Visitor Center are led by local astronomers such as CfDS member Martin Male, who photographs the night sky from the area (Fig. 3.36). Check out these websites for more information:

   theromneymarsh.net/visitors/centre.htm
   www.ryesussex.co.uk/accommodation.asp
   www.wadhurstastro.co.uk

2. HIGH WEALD AREA OF OUTSTANDING NATURAL BEAUTY. Adjacent to Romney Marsh is another place in the densely populated and generally light-polluted southeast of England where it is still possible to experience star-strewn skies.

Fig. 3.36 Comet Lovejoy from Romney Marsh, January 2015. (Photo: Martin Male)

The High Weald AONB, a nationally important and protected area of unspoiled terrain, described as "one of the best surviving, coherent medieval landscapes in Northern Europe," combines a good lighting policy with outreach activities. For example, the AONB partnership actively encourages schoolchildren, with the help of online resources for teachers, to observe the area's night sky visually and supplies night-sky meters. The website promotes local dark-skies awareness and supports an initiative in Wadhurst to achieve IDA Dark-Sky Community status. The best SQM measurement taken by local astronomer Phil Berry in the Wadhurst area of the AONB in 2012 was 21.09, confirming a remarkably dark night sky for this part of the country. Phil, who takes striking photos (Fig. 3.37) of the Milky Way from this area, recommends the Forestry Commission's Snape Wood car park, a very dark and isolated site towards the southern borders of Wadhurst. The narrow Brinkers Lane forest entry leads to a very small car park adjacent to a railway line. The Wadhurst Astronomical Society sometimes uses the Ashdown Forest, which has many large parking places such as Four Counties, Gill's Lap and Wren's Warren.

www.highweald.org
www.highweald.org/look-after/dark-skies.html
www.wadhurstastro.co.uk/WadhurstDarkSkyVillage2012.pdf

Finally, here is where to go if you are interested in visiting the southern islands of England.

Fig. 3.37 The Milky Way from Wadhurst. The prominent nebula right of center is NGC 7000, the North America Nebula. (Photo: Phil Berry)

1. SARK. Sark, a Channel Island protectorate of the United Kingdom near the coast of Normandy, became Europe's first International Dark-Sky Community, in 2011. Sark (including the nearby island of Brecqhou), with an area of just 2.10 square miles (5.44 km^2), is famed for its absence of cars; however, thousands of visitors arrive each year. No road lighting means that real darkness reigns here. Just about anywhere on Sark is an ideal place to savor the Milky Way, seen in great detail, and countless stars from horizon to horizon (Fig. 3.38).

Martin Morgan-Taylor, board member and past vice-president of the IDA, visited Sark in 2010 to verify the quality of the sky for the island's application for dark sky status. Survey work was carried out by Steve Owens, and the formation of the master plan was contracted to lighting consultant Jim Paterson. In 2016 Paterson was a winner of the British Astronomical Association's Joy Griffiths[4] Award for his sky-friendly lighting schemes; he wrote that "Sark's Chief Pleas (the island government) hope to influence other surrounding Channel Islands towards protecting the intrinsic darkness of their night-time environment."

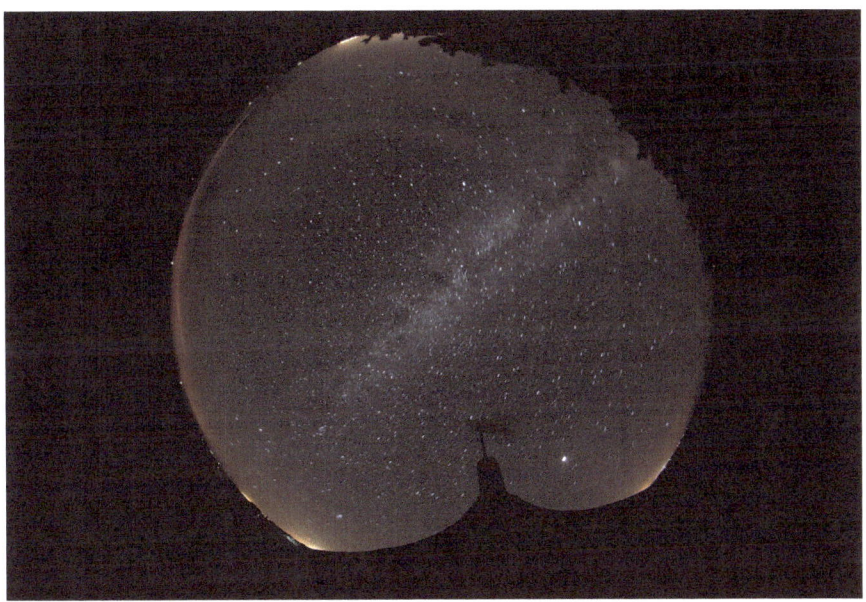

Fig. 3.38 The Milky Way bisects the sky over Sark. (Photo: Martin Morgan-Taylor)

[4] Joy Griffiths was one of the BAA Commission for Dark Skies' many local officers in the UK, and, despite a disabling condition limiting her movement, she effected great changes for the better in lighting in her area of southwest England. Following her sad and premature death, Commission members created the annual award in her name.

In October 2015, after much planning and fundraising, the Sark Astronomy Society (SAstroS), completed the island's first observatory.

www.sark.co.uk

darksky.org/idsp/communities/sark

2. ALDERNEY. On the island of Alderney, northernmost of the Channel Islands and within sight of the northwestern tip of the Cotentin peninsula in France, British Astronomical Association member and astrophotographer Michael Maunder organizes observing evenings and has a regular slot on local radio stations Quay FM and Radio Guernsey. Alderney is hoping to join its neighboring island of Sark in achieving IDA Dark-Sky status.

To publicize the island's starry aspirations, an eight-strong crew made a 60-mile catamaran journey from Weymouth, on the Dorset coast, to Alderney in 2014, navigating only by the stars. Michael explains why the constellations seem to have that extra luster from the island: "On Alderney you have extremely clean air coming across the Atlantic for 4,000 miles. Most of the island is near sea level, so the air hasn't got into turbulent motion, but is in laminar flow. In England laminar flow gives the effect of viewing through frosted windows, making the stars twinkle. Here we see objects that are very difficult to see there." Fort Tourgis, on the western flank of Alderney, is mentioned on local websites as one of many accessible observing places, and the island's relatively mild climate will be welcomed by less hardy observers.

www.alderneyholidays.com/2014/02/dark-sky-status-alderney
www.visitalderney.com/stay/accommodation

3. LUNDY ISLAND. Twelve miles off the Devon coast, the narrow, 3.5-mile long island of Lundy is the largest in the Bristol Channel. Lundy means 'Island of the Puffins.' This relatively treeless granite outcrop with its tall cliffs is easily seen from the mainland. It has a very small permanent population, and amenities are basic. In September 2014 Lundy was added to the West Country's growing list of Dark Sky Discovery Sites, and its airfield became the Discovery Site for the island. Lundy's few lights go off around midnight, and, with minimal light pollution from the mainland, there are spectacular views of the Milky Way and the constellations. The exposed landscape might suit hardier stargazers, who could stay at the island's campsite. Indoor accommodation consists mostly of a number of self-serve cottages.

Little Lundy has more than its fair share of birds and animals. Seabirds abound, though the puffin population, ravaged in the past by imported rats, is far less numerous than it once was, but is responding to conservation efforts. Sika Deer, feral goats and the somewhat primitive, very hardy Soay sheep roam the grassy landscape.

www.landmarktrust.org.uk/lundyisland

The size of the United States means that its climate is much more varied than Britain's. Climatic zones range from Mediterranean and continental through alpine to desert, and extremes of heat and cold are greater than those found in the UK.

A description of weather in the United States could fill a whole book, so the would-be traveler to any of the sites mentioned should consult websites such as www.ncdc.noaa.gov/customer-support/partnerships/regional-climate-centers.

The National Park Service has a Natural Sounds and Night Skies division that gives information about dark skies in national parks:

www.nightskyparks.org/home
www.nps.gov/orgs/1050/index.htm

For stargazing events in various states throughout the year, see:

www.amsky.com/calendar/events
www.astromax.org/links/starparty.htm

Annual frequencies of auroral displays in the United States are discussed on NASA's aurora website: pwg.gsfc.nasa.gov/polar/EPO/auroral_poster/aurora_all. pdf. In Alaska aurorae may be seen on most nights when skies are dark, perhaps ten times a year in far northern states, reducing to once or twice a year in southern states such as Texas and Florida at times of high solar activity.

Let's explore the 50 states one by one for their darkest skies, including details of some of the events held there in celebration of those skies.

Alabama

East of Brewton in the far south of Alabama is the 131-square-mile (340 km^2) Conecuh National Forest, where skies of Bortle Scale Class 2 can be found. The Auburn Astronomical Society (AAS—see website below) has produced maps of light pollution in central and southern Alabama, and gives details of its preferred dark site (including Bortle scale readings) in the Conecuh Forest. AAS member John Tatarchuk describes the view from this excellent site when skies are transparent: "The faint and elusive Gegenschein is visible every night, and on a good night, the incredibly faint zodiacal band (not to be confused with the immensely brighter zodiacal light) is visible, too. M33 (the Triangulum Galaxy) is about as easy to see with the naked eye as M31 (the Andromeda Galaxy) from a less dark magnitude 4.5 or 5 site. The Milky Way is incredible – the perfect Alabama dark sky site."

By day, the visitor can tour the shallow ponds, marshland and longleaf pine stands that typify the forest, but there is enough open and dry land to provide good sites and horizons for starwatchers. The Conecuh Trail is open all year, though websites recommend outdoor hikes during cooler parts of the year, "when insects are not so bothersome." The Open Pond Recreation area camping ground offers 'primitive' campsites and more developed ones with water and electricity.

The town of Andalusia, 10 miles north of the forest, is the birthplace of cosmologist Frank J. Tipler, well known for his contributions to debates on extraterrestrial life and the anthropic principle. Andalusia has several chain and smaller hotels.

www.auburnastro.org/trips/cnf/conecuh.htm

alltrails.com/parks/us/alabama/open-pond-recreation-area
hotelguides.com/alabama/andalusia-al-hotels.html

Alaska

Obviously, much of Alaska, with its vast uninhabited terrains, will be free of light pollution (Fig. 3.39). Accessibility, the climate in winter and the lack of night-time darkness in the summer are determining factors to take into account when assessing Alaska's suitability for starwatching. Southern Alaska, at the same latitude as northern Scotland, minimizes these factors.

The Eagle River runs to the north of Anchorage (61°N) and meanders into Eagle Bay, off the Cook Inlet of the Gulf of Alaska. The Eagle River Nature Center to the northeast of Anchorage hosts astronomy sessions on the first Friday of the month

Fig. 3.39 The aurora vertically above. (Photo: David Cartier, Sr.)

from October through April. Amateur astronomers from the community and staff from the University of Alaska at Anchorage's astronomy department present topics, followed by outdoor sky viewing with telescopes. Visitors are encouraged to bring their own instruments. About half of the time, says Center Director Asta Spurgis, the weather cooperates. Observers should dress very warmly as, reports Asta, "the best nights are when it's in the minus temperatures!" All lights are turned off during astronomy sessions. The center has cabin and yurt (tent) accommodations, and Anchorage has many hotels.

Further north, the University of Fairbanks, the only university in the world with its own rocket range, is famed for its research into and monitoring of the aurora. Fairbanks is the only American city typically beneath Earth's auroral oval. A search for "Alaska Aurora Tours" will reveal several organizations running northern lights experience trips.

Alaska Route 2 (the Richardson Highway) runs south from Fairbanks to Delta Junction and Alaska Route 4 continues from Delta Junction to Valdez. This is the state's oldest overland route begun in 1898 by the U.S. Army to supply Eagle Fort on the Yukon River. Incidentally, this was the way into the interior for many of the hopeful gold prospectors drawn into the Klondike Gold Rush. There is accommodation at Delta Junction, 96 miles south of Fairbanks, and 38 miles further south is the Lodge at Black Rapids, a place about which astrophotographers have said, "It's got it all: accessibility, glaciers, clean air and some of the best night skies we've seen," and "favorite place for the auroral displays."

www.ernc.org
www.ernc.org/river-yurt.html
www.alaska.org/destination/anchorage/hotels
www.explorefairbanks.com
www.gi.alaska.edu/AuroraForecast

Arizona

In Arizona, there's a lot of serious astronomy and casual stargazing going on! The state is renowned for its Meteor Crater, located just off Interstate 40 near Winslow in northern Arizona, and Flagstaff is known among astronomers for its observatory, open for visits, its dark sky ordinances, its many dark areas and the range of astronomical activities available. Tucson is said to have the darkest night sky of any city of comparable size in the United States, thanks to a dark sky ordinance passed in the early 1970s and updated in 2012. Flagstaff has similar protection for its night skies (www.tucsonnewsnow.com/story/17148223/pima-county-approves-updated-dark-skies-ordinance).

About 30 miles northeast of Tucson, and protected from any residual skyglow by the Santa Catalina Mountains, is Oracle State Park. The park was designated an International Dark Sky Park in 2014 by the IDA. It is a wildlife preserve, rich in the history of prehistoric peoples. The IDA describes it as "readily accessible to the nearly one million people who live in the Tucson metropolitan area… a convenient, affordable and authentic dark sky experience." Its relative remoteness means that it is one of the Arizona state parks' less visited sites, and potential night-sky observers are advised to check on its website for opening times.

In northwestern Arizona is the rugged Grand Canyon-Parashant National Monument, an extensive area of high plateaus to the north of the great canyon and recognized in 2014 by the IDA as a gold-tier Dark Sky Park (Parashant International Night Sky Province-Window to the Cosmos). Astronomers who have seen myriad stars over this challenging landscape report high frequency of cloudless skies and atmospheric clarity.

To the northeast of Phoenix, two resorts, the Four Seasons Resort Scottsdale at Troon North and Boulders Resort and Spa, offer stargazing programs for guests (Fig. 3.40). Tour company Stellar Adventures organizes stargazing trips in the Sonoran Desert.

  azstateparks.com/Parks/ORAC
  azstateparks.com/starparty
  darksky.org/idsp/finder
  www.nps.gov/para/planyourvisit/hours.htm

Fig. 3.40 Arizona star watch. (Photo courtesy Scottsdale Convention & Visitors Bureau)

www.experiencescottsdale.com/stargazing-under-the-sonoran-desert-sky
www.fourseasons.com/Scottsdale
www.theboulders.com

Arkansas

The best night skies in Arkansas are to be found in the west and northwest of the state.

Two hours' drive from Little Rock is Mount Magazine, Arkansas' highest point at 2753 feet. Mount Magazine state park's cabins and large mountain lodge offer, according to its website, "dramatic sunrises and sunsets, and… the spectacular sight of the star-filled night sky." The weather in this elevated area can change rapidly, with temperatures cooler than in the lower valleys. Low clouds and fog may be a problem, but when it's clear it is, in the words of a local astronomer, "a great place for starwatching."

Several state parks in Arkansas host astronomy events. Every other month the Sugar Creek Astronomical Society (SCAS, Bella Vista) offers free viewing nights at Hobbs, Arkansas' largest state park, in the southwestern Ozarks. Visitors may bring their own telescopes, though the SCAS invites observers to share their instrument, named "Big Boy." Other state parks promoting observation include Lake Ouchita, Lake Fort Smith, and Pinnacle Mountain, where the Central Arkansas Astronomical Society (CAAS, Little Rock) organizes star watches.

Members of the CAAS have recommended other locations, including, in northwest Arkansas: White Rock Mountain, Rich Mountain and Brock Creek Lake Recreation Area (Ozark National Forest); and in north-central Arkansas, Buffalo Point and Tyler Bend campground. The society's website lists upcoming night-sky observing events in its area.

www.mountmagazinestatepark.com
nwastargazers.org (Sugar Creek Astronomical Society)
www.friendsofhobbs.com/content/astronomy-nights
www.caasastro.org

California

From the Lassen Volcanic National Park in the northeast (Fig. 3.41, and see Chap. 1) down to Death Valley in the south, California has its share of starry vistas. British Astronomical Association member George Hurrion told of his encounter with coal-black skies near Lone Pine, California, on US395. A short drive away from the town, he enjoyed "an exceptional view of the Milky Way, not just as a band of stars but in the way it splits into different arms. It was fascinating to watch

Fig. 3.41 Stargazing at Manzanita Lake, Lassen Volcanic National Park. (Photo: Alison Taggart-Barone)

how the rotation of our planet made the Milky Way gradually move from one side of the sky to the other."

A hundred miles further to the east of Lone Pine along California Routes 136/190, skirting the northern shore of the now dried-up Owens Lake, is Death Valley National Park (www.takemytrip.com/09loneliest/08n_43a.htm).

In 2013 Death Valley's near-pristine nightscapes earned it the status of an IDA Dark-Sky Park—and it is the world's largest. The IDA comments that "the skies there are affected by only the smallest amounts of light pollution, classifying it at the highest level of IDA designation and star-filled skies, the gold tier. Astronomical objects seen there are available in only the darkest locations across the globe." Eureka Valley is a *dark* site among dark sites. Astrophotographer Dan Duriscoe of the National Park Service's Natural Sounds and Night Skies Division considers Eureka Valley to be "about as isolated as you can get in California" (quoted by Paul Bogard in *The End of Night*—see bibliography).

Death Valley's park rangers suggest that you "check out Mesquite Flat Sand Dunes or Dante's View for stargazing!" Dante's View is an elevated viewpoint terrace on the north side of Coffin Peak, along the crest of the Black Mountains, overlooking Death Valley. The rangers tell me that it has "the best view of Death Valley by day, and of its stars by night." This unspoiled wilderness has become a byword for extreme climate: be prepared.

Another British Astronomical Association member recalls a visit to California's Yosemite National Park, to the northwest of Death Valley: "Try Glacier Point:

Fig. 3.42 Orion and the Milky Way in a perfect California sky. (Photo: Nick Hart)

tremendous view across the valley during the day, and this must be the ideal place for seeing the Milky Way and vast numbers of stars at night" (Fig. 3.42). The park takes its role in preserving dark skies seriously (see website below). Local astronomy groups from the Bay Area and the Central Valley join the National Park Service to hold star gatherings in the summer at the Glacier Point Amphitheater. The historic Ahwahnee Four-Star Hotel stands on the valley floor within sight of Glacier Point.

In Inyo County in eastern California, the Bristlecone Pine Forest of White Mountain, the third highest peak in the state, is open from mid-May through the end of November, weather permitting. Not only does this area have the world's oldest trees, with one more than 5000 years old, but it also has good dark skies.

www.lonepinechamber.org
www.nps.gov/deva (Death Valley)
www.nps.gov/deva/planyourvisit/lodging.htm

Yosemite National Park Lighting Guidelines: www.nps.gov/yose/learn/nature/
   upload/Lighting-Guidlines-05062011.pdf
www.yosemitepark.com/Lodging

Colorado

The southern part of Colorado has some of its darkest skies (Fig. 3.43). Indeed, the towns of Westcliffe and Silver Cliff, between the Wet Mountains and the Sangre de Cristo Range in south-central Colorado, became the state's first IDA International Dark-Sky Community, in 2015. To the southwest is the Four Corners meeting place of the state lines of Colorado, Utah, New Mexico and Arizona. Here the crossing point is marked by a monument, whose metal ground markers resemble those at the Greenwich Observatory's Prime Meridian site in London. Nearby Cortez might be a good base for exploring the terrain and the stars of Hovenweep National

Fig. 3.43 The Milky Way from the lip of the Grand Canyon. (Photo: Nick Hart)

Monument, in southwestern Colorado and southeastern Utah between Cortez and Blanding, Utah. Both towns are well located for visiting several fascinating places in this part of the United States. In July 2014, the IDA granted Hovenweep, which means "deserted valley" in the native Western and Southern Numic languages, the status of an International Dark-Sky Park. The park takes pride in its "primordial dark sky, largely unaltered by modernity" and claims that "the night sky is about as dark as it was 800 years ago." Stargazing activities are confined to the Visitor Center parking lot and campground, and the park has taken measures to control any lighting in and around this area to minimize any impact upon the celestial wonders above. Some of the park's rock art and structures reflect aspects of ancient sky lore or events such as solstices.

Further east, the elevated basin of the San Luis Valley, from which the Rio Grande flows into New Mexico, has some "stunning" views of the stars, according to members of the British Astronomical Association. These are mostly desert lands, and bear in mind that they can be very cold at night. Alamosa is the valley's largest town.

darksky.org/idsp/communities/thecliffs
darksky.org/idsp/parks/hovenweep/
cityofalamosa.org

Connecticut

The Astronomical Society of New Haven (ASNH) freely admits that the relatively small state of Connecticut (Texas is nearly 50 times bigger) "is not one of your darker areas in the United States," but being typically resourceful and determined astronomers, its members have found a dark sky location in the least light-polluted area of the state, its northwestern corner. Near Goshen, home of the annual July Connecticut Agricultural Fair, the ASNH held its 25th anniversary Connecticut Star Party in October 2015. This took place at the Edmund D. Strang Scout Reservation, with three different designated observing locations offering "beautiful dark skies." You can drive from Hartford to the northwestern corner of the state in just over an hour. The Lichfield Hills website gives details of attractions, events and accommodation in the area.

Further east, campsites in Natchaug State Forest, known for its striking colors in autumn, and the Last Green Valley, a National Heritage corridor in eastern Connecticut and south-central Massachusetts, offer darker skies than can be found in the vicinity of the region's big cities such as Hartford, though they are not as dark as Connecticut's western reaches.

Under what may be the best night skies in the southwest of the state is the New Pond Farm Observatory in Fairfield County. Volunteers offer monthly public programs.

www.asnh.org
ctvisit.com/regions/details/litchfield-hills/3

thelastgreenvalley.org
www.newpondfarm.org/astronomy_observatory

Delaware

Looking at any of the U.S. light pollution maps listed might suggest that there's not much dark sky astronomy going on in Delaware. Stray light has no respect for state lines, and Atlantic City, Philadelphia, Baltimore and Washington, D.C. color the sky from great distances. However, organized stargazing programs and events are available in northern Delaware in several out-of-town state parks such as Bellevue, White Clay Creek and Brandywine Creek. In the southern part of the state there are even darker skies. At Holts Landing and Delaware Seashore state parks, telescopes are aimed skywards. Consult the state parks' websites (below) for details of access times and events.

Peter Graham, president of the Delmarva Stargazers, mentions good skies from Trap Pond State Park, near the town of Laurel, in the southeastern corner of the state and arguably its darkest area. The Stargazers have worked with the park to establish an observing area at Cypress Point, and a primary aspect of the collaboration has been the abatement of light pollution. SQM measurements of better than 21.5 (dark rural sky) are evidence that this is one of the best stargazing locations in Delaware. Annual star parties take place at Cypress Point, and the park allows camping at the observing location for a minimal fee. There are various RV and camping sites in the area with water and electricity (see camping website below).

www.destateparks.com/Activities/stargazing/index.asp
www.destateparks.com/camping/trap-pond/index.asp

delmarvastargazers.org/index.html

Florida

Some parts of Florida retain good dark skies. In the south, the South Florida Amateur Astronomers Association (contact via Observing Sites website, below) offers information on Everglades area astronomy. In central Florida, 30 miles north of Lake Okeechobee, are the 54,000 acres of the Kissimmee Prairie Preserve State Park, named for the river on its western edge rather than the town of Kissimmee away to the north. After exploring its varied birdlife, its grasslands and the largest remaining stretch of Florida dry prairie, you can discover the dark skies of one of Florida's premier stargazing destinations. Its remote location within the state means that light pollution is minimal, and it is reputed to be the darkest place in Florida. The Central Florida Astronomical Society has taken SQM readings of 22.12 (very dark indeed) here. In December 2015, the park celebrated the success of its bid for IDA Dark-Sky Reserve status. Find information about Kissimmee Prairie's dark

Fig. 3.44 From the AstroPad at Kissimmee in Florida. (Photo: Chris Clauson)

sky observing sites at the link below. Park manager Evan Hall states that "near our office we have five Observation Platforms located on our Astronomy Pad (Fig. 3.44). They are available for reservations 11 months in advance and are quite popular on the New Moon with astronomers from many of the local astronomical societies. Each pad consists of an area for a tent or small camping unit, with an adjacent area for telescopes and other viewing equipment."

Resident astronomy volunteers give daily presentations and talks, especially in the busier winter season, and local societies lead "Intro to Astronomy" programs.

The third and fourth links below list other state parks in Florida where stargazing activities take place, and accommodations.

www.observingsites.com/ds_fl.htm
www.floridastateparks.org/kissimmeeprairie
www.floridastateparks.org/activity-list/Star-Gazing
www.floridastateparks.org/things-to-do/stay-the-night

Georgia

In the southeastern corner of Georgia is the Okefenokee National Wildlife Refuge. Dark skies maps show this to be the state's starriest place, and they suggest that around the town of Fargo skies are the darkest of the dark. The Okefenokee Swamp

home page calls the Wildlife Refuge area "a black hole in a sea of light pollution…
the night sky is so clear, dark and open that it closely rivals a sky you may find in
western deserts… Great stargazing." Lights are controlled with timers and switches
within the refuge. You are invited when setting up your reservation or checking in
to discuss lighting where you will be staying.

The Georgia observing sites website mentions dark sky locations near LaGrange
and West Point in the west of the state, and near Oakman in the north. To the northeast of Oakman, Doll Mountain Campground, where amateur astronomers congre-
gate, has stars at the zenith down to magnitude 6, the faintest that the human eye
can normally see. It is open from early April until the end of October. The
campground has some lighting and, because stray light travels far, it is possible to
see the distant glow of the lights of Atlanta, 90 miles away to the south. In Taliaferro
County, one of the least developed areas in the east of the state, is Deerlick
Astronomy Village, near the town of Sharon. Here there are private lots used by
astronomers, with strict lighting rules. There is also a guest area for observing, with
hot water and electricity, bringing in astronomers from far and wide. Michael
Covington's photo (Fig. 3.45), taken from there, gives a flavor of Deerlick's stellar
attractions.

www.fws.gov/refuge/okefenokee
www.okefenokee.com/accomodations/okefenokee-cabin-rentals
www.okefenokee.com/nature_ecology_of_okefenokee_swamp/stargazing

Fig. 3.45 In Georgia skies, dark nebulae surround the Sagittarius star cloud. (Photo: Michael
Covington)

www.observingsites.com/ds_ga.htm
www.recreation.gov/camping/doll-mt-campground
www.ajc.com/news/news/local/atlanta-star-gazers-head-to-east-georgia-
    for-darkn

Hawaii

The islands of the Hawaiian chain are washed by the waters of the Pacific, where the darkest ocean skies on the planet are to be found. Astrophotographer Carey Johnson, who lives on Oahu, would be one of the first to admit, though, that not all Hawaiian islands are the best places for starwatchers. "It's pretty difficult on Oahu because of light pollution. There are some dark places (Fig. 3.46) at the west tip of the island, but it would be much darker on every one of the others: Kauai, Molokai, Maui, and the Big Island of Hawaii, with its telescopes at 14,000 feet."

On the Big Island, 13 international observatories cluster around the summit of Mauna Kea, a mighty and fortunately dormant volcano. Local anti-light-pollution legislation protects the observatories and the surrounding area. Thousands visit Mauna Kea every year, but the mountain's elevation raises concerns over altitude sickness and weather for those who wish to discover its summit.

Fig. 3.46 Ghostly Comet Holmes (2007) and the California Nebula in Perseus, taken from Oahu's north shore. (Photo: Carey Johnson)

Lower down, and on every night of the year, the Visitor Information Station (VIS) at 9200 feet presents a free nightly stargazing program from 6 pm until 10 pm. Reservations are not needed. There are telescope piers at the VIS, and local astronomers hold star parties.

The more intrepid astro-tourist who would like to see the stars from higher up might care to search "Mauna Kea Summit Tours."

There is a considerable range of hotels and guest houses around the coasts of the Big Island. See www.gohawaii.com/en/big-island.

   www.ifa.hawaii.edu/info/vis/visiting-mauna-kea/star-gazing-program.html
   www.ifa.hawaii.edu/info/vis/uploads/images/article.jpg

Idaho

Away from Boise, most of Idaho has either very acceptable or extremely dark skies at night, as a look at darksitefinder.com/maps will confirm. The wonderfully help-ful staff of the Craters of the Moon National Monument and Preserve, 18 miles southwest of Arco, southern Idaho, provided a wealth of information and images of this unusual and intriguing place (Fig. 3.47). Ranger Jack Davisson states: "We have a great night sky here at Craters of the Moon. A good place to take in the view

Fig. 3.47 The Milky Way over cinder cones at the Craters of the Moon National Monument, Idaho. (Photo: Daniel George)

would be on top of Inferno Cone (one of our accessible cinder cones). The Idaho Falls Astronomical Society hosts a star party in our Cave Area parking lot every June. It's a pretty amazing event."

To the west of the Craters of the Moon is Bruneau Dunes State Park, south of Mountain Home in Idaho's southwestern corner. It is an area of large sand dunes and a small lake, but the main astronomical attraction is the Bruneau Dunes Observatory (Fig. 3.48), open from early April through mid-October on Fridays and Saturdays. In the daytime, the observatory offers free tours of the facility and visitors can also study the Sun through a special solar telescope. On clear nights and for a small fee (five-year-olds and under free) there is skywatching using the observatory's telescopes.

www.nps.gov/crmo/learn/nature/night-sky.htm
www.nps.gov/crmo/learn/news/star-party.htm

parksandrecreation.idaho.gov/parks/bruneau-dunes
visitidaho.org/places-to-stay

Fig. 3.48 The Bruneau Dunes Observatory. (Photo courtesy Bruneau Dunes State Park)

Illinois

In November 2011 the town of Homer Glen, 30 miles southwest of Chicago, became an IDA International Dark-Sky Community. Officials in Homer Glen, mindful of its relative proximity to the Windy (and big!) City, promoted a "smart lighting" policy, bolstering dark sky awareness and trying to ensure that its skies were not tainted by local lights. The Illinois Coalition for Responsible Outdoor Lighting (ICROL), founded by concerned residents, successfully called for a state-wide resolution on good exterior lighting, and this was adopted in 2010. Local astronomy groups in the Homer Glen area arrange the Biannual Stargazing Events program at Trantina Farm. Amateur astronomers from the Kankakee Area Stargazers, the Naperville Astronomical Association and the South West Astronomy Observers Group guide visitors around the cosmos.

Even darker skies and organized star watches are to be found further south at the Jim Edgar Panther Creek State Fish and Wildlife Park, about 30 miles northwest of Springfield and home of the annual Illinois Dark Skies Star Party, presented by the Sangamon Astronomical Society. The event includes an equipment swap, talks by guest speakers and an astrophotography contest.

The northwestern corner of the state has some sparkling skyscapes—base yourself in or near Mississippi Palisades State Park near Savanna and explore the area for likely observing sites.

Clyde Tombaugh, who famously discovered Pluto in 1930, was born in Streator, Illinois, in 1906.

> www.homerglenil.org/2413/Stargazing
> www.illinoislighting.org
> www.skyandtelescope.com/astronomy-events/illinois-dark-skies-star-party-2

dnr.state.il.us/lands/landmgt/parks/r1/palisade.htmweekendsightseer.com/the-last-dark-skies-of-northwest-illinois

Indiana

IDA members Alan and Rosemary Bell were the moving spirits behind the recognition in 2014 of Beverly Shores, Indiana (Fig. 3.49), as an IDA Dark-Sky Community. Refits and replacements of old, wasteful "cobra-head" streetlights and a public commitment to conserving the night sky brought back the stars. Rosemary Bell tells me that "we no longer have skyglow in Beverly Shores. That, coupled with the fact that we are surrounded on three sides by the Indiana Dunes National Lake Shore and on the remaining side by Lake Michigan, allows excellent access to the night sky. The best spot for viewing the night sky is at the Indiana Dunes State Park parking lot, adjacent to Beverly Shores, which is light-free."

Fig. 3.49 Beverly Shores lakefront in winter. (Photo courtesy Alan and Rosemary Bell)

Between Louisville (Kentucky) and Evansville in southern Indiana is the Hoosier National Forest, one of the state's darkest areas. The Louisville Astronomical Society holds meetings at the James G. Baker Center for Astronomy, Curby, to the east of the forest. Contact the society for details. The darkest sites in this area are southwest of Leavenworth, on the Ohio River, between the Hoosier National Forest and the Harrison-Crawford State Forest. Mike Lynch's book *Indiana StarWatch* (ISBN 9781610604130) will be of interest.

darksky.org/beverly-shores
www.fs.usda.gov/attmain/hoosier/specialplaces
louisville-astro.org
www.gotolouisville.com/stay/index.aspx

Iowa

On U.S. Route 65 in Wayne County, southern Iowa, is Lineville, taking its name from its position on the state line with Missouri. Dark-skies maps show this area to be the darkest in the state. Just north of Lineville on the eastern side of Route 65,

Moore-Gosch Memorial Park with its large lake is a good place to see stars. It has several campsites with water and electricity. Check on opening times.

The Des Moines Astronomical Society (DMAS) runs a two-dome observatory at Ashton-Wildwood County Park in Jasper County, near Highway 330 to the northeast of Des Moines. The site has a spacious classroom and two 16-inch telescopes. From April through October there are free weekly public observing nights. Private group visits can be arranged.

At Whiterock Conservancy, near Coon Rapids on the Middle Raccoon River 70 road-miles west of Des Moines, the Ames Area Amateur Astronomers host the annual Iowa Star Party in August. There are talks and telescope sharing. See full details on the star party's website.

> www.mycountyparks.com/county/Wayne/Park/Moore-Gosch-Park.aspx
> dmastronomy.com/group-visits
> www.hotelplanner.com/zip/Coon-Rapids
> www.iowastarparty.com

Kansas

The Astronomical Society of Kansas City operates the Powell Observatory in Louisburg, built in 1985 by members with the support of a grant from the Powell family of Kansas City. It is open for group reservations and Saturday night "Star Bright" programs. Guided tours of interesting night-sky targets are provided at the eyepieces of the 30-inch Ruisinger Telescope and other instruments. For further information on opening times, fees and other details, see the program schedule on the Powell Observatory website below.

In areas of Kansas well away from Kansas City and Wichita, there are plenty of roadside places between its numerous small towns where you might stop and see a starry sky, and with a modest telescope examine some of the more accessible galaxies (Fig. 3.50).

A trip along route I-70 in northwest Kansas puts you in one of its darkest areas. Base yourself in Colby, Oakley or the larger Hays, or further west towards Wilson with its lake and state park. Drive southeast from Oakley towards Gove and explore the area known as Monument Rocks, sometimes called the Chalk Pyramids and a National Natural Landmark—a good dark sky site according to local astronomers. The area is open to the public and is approached from the highway along a gravel track. These 70-foot chalk formations were laid down 80 million years ago in what was then an extensive inland sea. From a distance they can look eerily like a natural Stonehenge.

Words of warning: in this and other predominantly agricultural states, farm lights can be a problem for observers in rural areas. Also, Kansas can be a windy place. Astronomers complain about light pollution hindering their observations, but

Fig. 3.50 NGC 4565, a needle-shaped galaxy in Andromeda, accessible with a modest telescope in dark skies. It is 40 million light years away. (Photo: Paul and Liz Downing)

wind can be almost as irritating, as discomfort hinders concentration when using telescopes, and a very solid mount is needed to minimize vibration.

    www.askc.org/powell.htm
    www.northwestkansas.org/counties
    www.kansastravel.org/monumentrocks.htm
    www.travelks.com/lodging

Kentucky

Some of the darkest sites in the northern part of Kentucky are south of the Ohio River near the state line with Indiana, across the water from areas mentioned in the Indiana entry. However, the light-pollution map at www.jshine.net/astronomy/dark_sky shows the state's best skies to lie in a crescent in eastern Kentucky, the northern tip being around the aptly named town of Moon, a small community in Morgan County, along Route 172 northeast of West Liberty. This crescent of comparative darkness continues westwards through the Sandy Hook area, then turns southwards around West Liberty, along the northern edge of the Cumberland

Plateau. It then plummets south toward the Cumberland Gap and Middlesboro, in Bell County.

Two sites with notable dark skies along this crescent are the Chimney Top Trail and the Redbird Wildlife Management Area. Check accessibility and opening times for both.

Chimney Top Trail (Trail #235), leading to superb daytime views from Chimney Top Rock, is located in the pine-hardwood forests of the Red River Gorge in the Cumberland District of the vast Daniel Boone National Forest. Proceed with care! Semi-primitive forest camping is available at the Koomer Ridge Campground, with access to the trails in the Red River Gorge.

The 25,000-acre Redbird Wildlife Management Area is located in Leslie and Clay counties, in the Redbird District of the Daniel Boone National Forest, and is popular with hunters, anglers and hikers as well as astronomers.

> www.fs.usda.gov/recarea/dbnf/recreation/recarea/?recid=39526&actid=50
>   (Chimney Top Trail)
> www.kentuckytourism.com/koomer-ridge-campground
> www.kentuckytourism.com/redbird-wildlife-management-area

Louisiana

On dark moonless night the astro-tourist seeking a Louisiana starscape might well visit the Catahoula National Wildlife Refuge. The publicly accessible refuge is situated in the eastern-central part of the state, about 20 miles/32 km to the east of Pineville. Explore the area in daylight for the best pull-offs and parking areas with low horizons. Artificial light is not too obvious here, and the only downside might be biting insects around the lake at certain times of the year.

Louisiana's star party is organized by the Louisiana Office of State Parks, the A. J. and Nona Trigg Hodges Foundation and the Baton Rouge Astronomical Society. The Hodges Gardens Star Party welcomes visitors annually at its observation site in the Hodges Gardens State Park in west-central Louisiana. Its website tells us that "in addition to a wonderfully dark night sky, the park offers opportunities for fishing, canoeing, birding, hiking and nature photography. It's become our tradition to invite the general public to view the sky with us on Saturday night." Basic camping and RV parking are allowed in designated areas, and the park has lakeside cabins for hire.

> www.fws.gov/refuge/catahoula
> www.brastro.org/hgsp.html
> www.louisianatravel.com/places-to-stay

Maine

A great arc of night-time darkness curls around the northern half of the state of Maine. At the center of this arc is Mount Katahdin, surrounded by the North Maine Woods, hundreds of thousands of acres of forested land, lakes and streams. Not so long ago the area was owned by lumber companies, and much of it remains unchanged since Thoreau's exploration of it in the 1840s. Above the North Maine Woods and Downeast Maine are some of the largest areas of dark sky east of the Mississippi River.

In 2014 the first annual Stars Over Katahdin (SOK) event was held, with the intention of it being a yearly program around the September new Moon. Katahdin Woods and Waters (KWW) Recreation Area, a unique terrain of 100,000 acres of pristine wilderness and primordial night skies to the east of Mount Katahdin and Baxter State Park, offered to co-sponsor this event along with the Maine Chapter of the International Appalachian Trail, which has its beginning on this land.

Bar Harbor, on Mount Desert Island, Maine's largest offshore island, is at the eastern tip of the state's swathe of darkness. The town, near Acadia National Park, passed lighting ordinances in 2009 to protect starry night skies. That same year, the first Acadia Night Sky Festival was held as a result of this action.

Surrounded by ocean, Acadia National Park has good dark skies. This 47,000-acre terrain of forest and rock-strewn coastline, with glacier-scoured granite peaks such as Cadillac Mountain, holds regular ranger-led star watches within its boundaries, for example at Sand Beach on the east side of the island. An interesting and unusual way to go stargazing is to book a night-time kayak tour out of Bar Harbor (www.acadiaparkkayak.com).

> katahdinwoods.org/visit
> www.barharborinfo.com/places-to-stay.aspx
> www.acadianightskyfestival.com
> www.themainemag.com/play/a-list/2743-stargazing-lookouts.html
> www.internationalatmaine.com

Maryland

The 50,000-acre Green Ridge State Forest (GRSF) near the Potomac River in the Appalachian Mountains of northwestern Maryland is far enough away from population centers to have retained starry skies. It is the largest contiguous block of public land in the state and has a good network of fire roads and nearly 100 primitive campsites. The website gives information on registering with GRSF headquarters, negotiating its trails and, importantly, "Tips for camping in black bear country." The marshland area around the Blackwater National Wildlife Refuge, 12 miles south of Cambridge, is another of Maryland's relatively dark places. The refuge itself opens only from dawn to dusk, but after spending the day there you might like to stargaze from nearby locations. Information on nearby lodging and camping is at the FAQ website below. Maryland has many astronomy groups. Look on the Go-astronomy webpage below for contact details and advice on dark sites and astronomy events.

> dnr2.maryland.gov/forests/Pages/publiclands/western_greenridgeforest.
>   aspx
> www.fws.gov/refuge/Blackwater/about/brochures.html
> www.friendsofblackwater.org/faq.html#lodging
> www.go-astronomy.com/astro-clubs-state.php?State=MD

Massachusetts

Waste light from Boston and adjoining urban areas, and from Springfield and Albany (NY), makes truly dark skies a thing of the past in much of the state of Massachusetts. In the darker western part of the state is the Arunah Hill Natural Science Center with some good vistas of the stars (www.observingsites.com/ds_ma.htm), and on Nantucket Island and Cape Cod, skies are better than average. The Cape Cod Astronomical Society has its own well-equipped Werner Schmidt Observatory in the grounds of the Dennis Yarmouth High School in South Yarmouth, and it can be visited by arrangement combined with free Thursday evening astronomy events.

The Maria Mitchell Association's Loines Observatory on Nantucket, "one of the best stargazing spots on the East Coast," hosts visitors on three nights a week. The observatory uses two telescopes, an antique 8-inch refractor dating from 1908, and a modern 24-inch research reflector. See its website for details of public tours, programs, lectures and special events. Maria Mitchell, born in Nantucket in 1818, was the first American woman to become a professional astronomer. Another renowned woman astronomer was Henrietta Swan Leavitt, born in Cambridge, Massachusetts, in 1868, whose work on variable stars underpinned later research by Hubble and others on the expansion of the universe.

> www.massvacation.com/navigate/travel-guide
> www.ccas.ws/newstarpartypage.html
> www.mariamitchell.org/visit/loines-observatory

Michigan

At the northwestern tip of the Lower Peninsula of Michigan is the Headlands International Dark-Sky Park, which gained IDA recognition in 2011. The park's caretakers describe programs "dedicated to a humanities-based approach to the sky and its stories, and the direct encounter between human beings and the natural environment." Headlands reports "tens of thousands of visitors" annually enjoying its starry skies. Beth Eckerle, Emmet County Director of Communications, working in conjunction with Dark-Sky Park program director Mary Stewart Adams, tells of the great efforts by park staff and Emmet county leaders to create a night-time environment where protecting the dark sky is a priority.

Artificial lighting is inconspicuous within the 243-hectare park, and Emmet county aims to keep the area as natural as possible. A county-wide outdoor lighting ordinance helps curtail the proliferation of artificial light at night in surrounding areas, while the land in and around the park is zoned for natural conditions with strict limits on outdoor lighting. The park hosts regular sessions celebrating the night sky, including night-time storytelling, starry cruises, and exploratory outings such as hikes and snowshoe adventures. There are astrophotography lessons, and special events including lunar eclipse and meteor viewings. There is an extensive dark sky observation area along the Lake Michigan shoreline, "a quarter of a mile's worth of spectacular viewing."

The Hiawatha National Forest in Michigan's Upper Peninsula also holds stargazing events.

The state's tourist websites carry extensive information about places to stay in the peninsulas.

> www.midarkskypark.org
> www.fs.usda.gov/hiawatha
> www.michigan.org/hotels-motels
> www.michigan.org/campgrounds-rv-parks

Minnesota

Stray light from Minneapolis, its surrounding urban areas and other towns dominates the night sky in much of the Minnesota heartland. In the State of Ten Thousand Lakes, darker areas are largely confined to a northern strip near the Canadian border and in areas close to Lake Superior. The Chequamegon-Nicolet National Forest (CNNF), 185 road-miles northeast of Minneapolis, retains its dark skies, though if you prefer more open horizons (Fig. 3.51) drive north or northwest from Wausau, east of Minneapolis. En route, there is much (long ago deforested) open farmland where starry vistas may be seen—the further north and northwest the better.

To the west of Duluth, the Savanna State Forest, near Libby on Minnesota State Highway 65, is a long east-west strip of darkness. To the south is Palisade, on Aitkin County Road 10. Near here, the Long Lake Conservation Center (LLCC) collaborates with the Minnesota Astronomical Society to stage the annual Northern Nights Star Fest in August. Large telescopes are in use, together with astro-themed events such as swap sessions and talks. Cabins and camping are available—book well in advance.

Fig. 3.51 The benefit of a low horizon: red Antares, in Scorpius, rises reflected in the lake. (Photo: James Hilder)

www.fs.usda.gov/cnnf
www.thisweeklibby.com/Where-to-Stay
www.trails.com/palisade_minnesota-hotels.html
www.mnastro.org/NNSF
www.longlakecc.org

Mississippi

Some of the best night skies in Mississippi are around Stewart and French Camp, in central Mississippi about 100 miles to the northeast of Jackson. Rainwater Observatory and Planetarium, on State Highway 413 nearby, has details on its website of free public astronomy events on the second Friday of each month, with indoor presentations and outdoor observing (if clear). The cloudy-weather alternative is a planetarium presentation.

Nearby Kosciusko to the southwest of French Camp might be the place to stay if you prefer hotels. In southern Mississippi the campgrounds of the De Soto National Forest, under relatively dark skies, are also worth investigating, or you could base yourself in Hattiesburg just to the northwest. The forest is crossed by several National Recreation Trails such as the Black Creek and the Tuxachanie, suitable for hikers or cyclists.

rainwaterobservatory.org/rainwater/index.cfm/information
kosciusko.ms/tourism
www.ohranger.com/desoto-forest
www.hattiesburg.org/index.cfm/stay/hotels-motels   www.wilderness.net/
    printFactSheet.cfm?WID=54 (Black Creek)

Missouri

Fifteen miles to the west of Butler, Missouri, is the designated dark sky site of the
Astronomical Society of Kansas City (ASKC, see also the Kansas entry). The lively
Heart of America Star Party (HOASP) happens annually in September at this location, which has been shown to have a limiting magnitude (faintest star seen) of
6.7 — a good dark sky (Fig. 3.52). Solar observing is on offer during the day. The
site is carefully tailored to visiting observers, with spaces for tents, trailers, and
RVs. There are toilets and showers, and electricity hook-ups for motorized telescopes. One-night or longer stays can be booked. Sharing telescopes is the norm,
and there are talks, competitions (ever played Astro Bingo?) and local trips.

A darker though more remote area of the state that may appeal to hardier starwatchers is the Mark Twain National Forest in south-central Missouri. Campsites

Fig. 3.52 Dark skies are needed to appreciate M27, the Dumbbell Nebula in Vulpecula. (Photo:
Neill Mitchell)

here are popular with astronomers, and are listed at the MTNF website. You might like to visit the (replica!) Hubble Space Telescope in nearby Marshfield, Edwin Hubble's birthplace not far northeast of Springfield on Interstate Highway 44.

www.askc.org
www.askc.org/powell.htm
hoasp.org
www.fs.usda.gov/activity/mtnf/recreation/camping-cabins

Montana

Would-be star watchers studying a map of Montana, sometimes called the "Big Sky State," might find their attention drawn to a place actually called Big Sky—the "Gateway to the Yellowstone National Park." It all sounds promising, and indeed, this corner of the state certainly has its dark skies; but the development since the 1970s of Big Sky into a premier national winter sports resort means that it may be a better place for skiing than for stargazing—though if you're into both, there's plenty of accommodation. The Big Sky Astronomy Club (BSAC) can advise on the best local observing sites. Roads in the area such as US Highway 191 leading down to the West Entrance to Yellowstone, and 287 running parallel to it through Ennis, offer places to stop and contemplate the heavens on clear nights. The most extensive swathes of darkness in the state, however, are to be found in its western and central parts. In the west, the BSAC Montana Starwatch takes place in summer, on a site near Twin Bridges, southeast of Butte. Its website mentions sky meter readings of 21.

In central Montana around the Upper Missouri River is a very large area of good-quality observing, the foci of which are the Upper Missouri River Break National Monument and the Lewis and Clark National Forest. The Lewis and Clark Interpretive Center teams up with the Central Montana Astronomy Society (CMAS) for various astronomical events, details of which can be found on relevant websites.

www.bigskyastroclub.org
montanastarwatch.org
www.lewisandclarkfoundation.org
www.cmasweb.com/prod01.htm

Nebraska

Away from the urban areas of eastern Nebraska, the north-central part of the state has its best night skies. The state's darkest terrains are to be found in an area of about a 100-mile radius around Mullen. In the north of this area, astronomers enjoy the annual Nebraska Star Party (NSP) at the Snake Campground, Merritt Reservoir, on a sandy plateau at an altitude of around 3000 feet, 30 miles south of Valentine. It's reported that, when there is no Moon, you can see your shadow cast by faint light from the Milky Way. The NSP is very beginner- and child-friendly, with special basic astronomy programs and events. Its contingent of advanced observers is ready to share telescopes and offer advice and guidance through extremely dark starfields. Primitive camping only is available at the observing site. The Merritt Trading Post on Highway 97 is 10 minutes' drive north of the site, with groceries, camping requisites, public showers and laundry facilities. The nearby Cedar Bay Campground offers electricity hook-ups and shower facilities.

www.nebraskastarparty.org
visitnebraska.com

Nevada

Nevada retains some wonderful night scenes (Fig. 3.53). Paul Bogard (see California entry) suggests the stunning skies of the Black Rock Desert of Nevada. The Black Rock Desert/High Rock Canyon Emigrant Trails National Conservation Area is in northern Nevada, far enough away from Las Vegas to have retained pristine night skies. Little human light is seen in this sparsely populated area, and

Fig. 3.53 The Milky Way in desert skies. (Photo: Nick Hart)

frequent cloudless skies mean horizon-to-horizon starfields. British Astronomical Association members who have visited the area recommend the views from the middle of the Black Rock Desert Playa, away from mountains and therefore with lower horizons. They also have fond memories of the Milky Way on moonless nights from the Wheeler Peak Scenic Drive in Great Basin National Park, in eastern Nevada near the state line with Utah.

Great Basin National Park features astronomy as an all-year activity. Organized summer astronomy programs are offered at the Lehman Caves Visitor Center, and telescopes are available. Astronomy activities take place both by day and by night; the park provides solar viewing opportunities for all ages using specialized solar telescopes, and summer full Moon hikes. The park's telescopes are funded by the Great Basin National Park Foundation. See the website for event start times and all other details. There are many places within Great Basin National Park, far from the madding lights of civilization, where 'lone wolves' can stargaze at their leisure on clear, moonless nights. The park recommends Mather Overlook and the Baker Archaeological Site, just outside the town of Baker—panoramic daytime views and big, star-filled skies.

The park staff runs an annual autumn Astronomy Festival, with many telescopes (some very large indeed) at hand. Solar observing during the day is followed by the reputedly not-to-be-missed Ranger Talent Show (!) and, if clear, observing the heavens after dark with the help of expert volunteer astronomers. Children earn their Deep Space Certificates (and a Milky Way candy bar) as well as viewing what many perhaps can never hope to see from their urban back yards. There is also a Night Sky Photography workshop.

The area around Lake Tahoe, straddling the Nevada/California border west of Carson City, has many fine stargazing sites. For a complete and tempting list see tahoequarterly.com/Article/lake-tahoes-best-places-to-stargaze.

> blackrockdesert.org/stargazing
> www.nps.gov/grba/planyourvisit/great-basin-night-sky.htm
> www.nps.gov/grba/planyourvisit/wheeler-peak-scenic-drive.htm

New Hampshire

A light-pollution map of the northeastern United States shows a finger of real darkness protruding from Maine into New Hampshire's Great North Woods region; within the tip of that finger is Errol, north of the White Mountains along Route 16 at the intersection of Route 26. Errol, on the Bear Brook, is situated to the west of Umbagog Lake, which straddles the border with Maine. The town hosts the Umbagog Wildlife Festival in August. The Great North Woods region, at the northern tip of the state, is a rugged place of lakes and forests attracting outdoor adventurers, and its night skies are well known to northeastern astronomers. The Dead Diamond River Valley to the north of Errol is arguably the darkest place in New

Hampshire. Plan your trip with care; rough roads and plentiful wildlife, including moose, are likely to be encountered here. There are lodges, campsites and small hotels in the Great North Woods area. Its larger towns are Colebrook, Columbia, and Pittsburg, all on U.S. Route 3.

The Saint-Gaudens National Historic Site in Cornish, southwestern New Hampshire, hosts a free one-night star party every year, usually in late September or early October. See the website below.

Not far to the north of the state across the international border are the observatories and guaranteed dark skies of Québec's Mont Mégantic, the world's first International Dark-Sky Reserve.

> www.homeaway.com/vacation-rentals/new-hampshire/errol/r6402
> www.visit-newhampshire.com/greatnorthwoods/hotels-and-motels
> www.ucampnh.com/great-north-woods-region.php (campsites)
> www.allroutes.to/thestars (Dead Diamond River)
> www.nps.gov/saga/planyourvisit/star-gazing-party.htm (Saint-Gaudens)

New Jersey

Light spill from Philadelphia, Atlantic City and other urban areas confine the state's best stargazing to pockets of relatively dark skies (usually reported at Bortle Class 4, rural/suburban transition) in its southern half. Examples are Wharton State Forest's Godfrey Bridge campground and the Atsion recreation area.

Further north in the state, in Voorhees State Park near High Bridge, is the Edwin E. Aldrin[5] Astronomical Center, operated by the New Jersey Astronomical Association (NJAA). The Center's Paul Robinson Observatory, at an altitude of 840 feet (256 m) is owned by the NJAA and possesses the largest telescope in the state. Smaller instruments are in use, including a solar 'scope. Skies here are rated at Bortle Class 4/5. See NJAA's website for opening times and public sessions.

> www.state.nj.us/dep/parksandforests/parks/wharton.html
> www.state.nj.us/dep/parksandforests/parks/voorhees.html
> www.xactstudios.com/njaa/index.html

New Mexico

Much of New Mexico has what the state's official website calls "enchanted skies" (Fig. 3.54). The Magdalena Chamber of Commerce hosts the Enchanted Skies Star Parties at a dark sky observing location in the Cibola National Forest, near Magdalena in central New Mexico. Not far from here is the National Radio

---

[5] Lunar astronaut "Buzz" Aldrin was born in January 1930 in Montclair, New Jersey.

Fig. 3.54 Enchanted skies: Venus rises over an arid landscape. (Photo: Nick Hart)

Astronomy Observatory (NRAO), which operates a number of facilities, including the Robert C. Byrd Green Bank Telescope, the world's largest fully steerable radio astronomy dish. Visitors are welcome at the NRAO.

In the northeastern corner of the state and 15 miles to the northwest of the town of Clayton is Clayton Lake State Park, which celebrated the creation of its Dark-Sky Park in 2010. SQM values of 21.6 have been recorded. Local high school principal Terrell Jones pioneered astronomical activities in the area and the enthusiastic Clayton Astronomy Club operates Star Point Observatory. The club's unofficial headquarters is Art's Barber Shop at 2 Main Street in downtown Clayton. Contact the president, Art Grine (claytonastronomyclub@gmail.com), or the Clayton Chamber of Commerce for information about club events and observatory visits.

Clayton is very dedicated to its astronomy activities, which are promoted in its streets, tourist guides and hotels. The Holiday Motel on Highway 64 to the northwest of the town has good reviews.

In 2013, Chaco Culture National Historical Park, in northwestern New Mexico, became a gold-tier level International Dark-Sky Park. The Albuquerque Astronomical Society collaborates with its star watches. The designation of over 99 % of its area as a "natural darkness zone" means that Chaco protects its nocturnal wildlife while welcoming those who seek out its stars. The park is, in its own words, "one of the best places in the country to experience and enjoy natural darkness." With frequent dark-skies events, a public observatory, strict lighting guidelines and educational programs, the park's astronomical profile is firmly established.

There have been good reports also of dark skies around Carrizozo, not far north of Alamogordo.

> enchantedskies.org
> www.gb.nrao.edu/epo/gp/
> claytonnm.org (Chamber of Commerce)
> www.nps.gov/chcu/planyourvisit/nightsky.htm (Chaco)

New York

Some, especially in the UK, imagine that New York State must be flooded with stray light from the urban spread of New York City and its environs, but the state is bigger than they think. Its area is greater than that of several European countries; it is marginally larger than Greece. The distance from Staten Island in the south to the Canadian border in the north is about 350 miles (560 km), and the northern part of the state has some excellent dark skies. Find them in the 6-million-acre Adirondack Park, with more than 10,000 lakes and 30,000 miles of rivers and streams. The park is heavily forested, though there are more open areas where horizons are lower. Try the Adirondack Public Observatory near Little Wolf Pond, a fairly dark site. The observatory is open for public viewing on the first and third Fridays of each month until late May. Private viewing sessions can be arranged.

Adventurous astronomers recommend in New York, among many other dark sites, Meacham Lake, Lewey Lake, Wakely Dam on the Cedar River Flow Lake, and the area west of Tupper Lake around Cranberry Lake State Campground, where night skies have been described as "spectacular." Be aware that bears may be encountered in the park, and that phone and GPS signals are likely to be non-existent at many sites. In clearly signed designated wilderness areas no motor vehicles are permitted.

There are many hotels and motels in the Adirondack region, though some are in the more light-polluted areas around Lake Placid and Tupper Lake. The Adirondack hotels website below offers a good selection.

> visitadirondacks.com/about/adirondack-park
> www.apobservatory.org
> www.adirondack.net/adk/hotels.cfm

North Carolina

One of the smaller parks in the IDA Dark-Sky Places program is to be found in the Blue Ridge Mountains of western North Carolina. Here, the Blue Ridge Astronomy Group (BRAG, Burnsville, NC) and Yancey County have established a publicly accessible 6-acre star park. A dedicated public observatory with a large 34-inch telescope is planned, and BRAG member Bob Hampton predicts that construction will begin in the spring of 2016. The site, just off Highway 80 N in Yancey County (between the towns of Burnsville and Spruce Pine) is shared with Energy Xchange, a waste methane-gas utilization project. The park, in the mountainous surroundings of the Pisgah National Forest, was the first such site in the southeastern United States to receive the IDA's Dark-Sky Park designation. Contact BRAG for updates on this project.

Further to the west, a crescent of night-time darkness crosses the Great Smoky Mountains National Park, which straddles the border between the western tip of North Carolina and Tennessee. Higher areas or peaks such as Clingman's Dome have superb daytime views and observation facilities, though you are more likely to see light spill from distant urban areas there at night. Lower-lying areas in the dark sky zone with wooded horizons suffer less residual light creep from towns. Examples are long, winding Fontana Lake to the west of Bryson City, or Lake Santeetlah on U.S. Route 129. If you approach the latter on 129 from the north, take great care on the difficult 'Tail of the Dragon' section, with its 318 curves in 11 miles.

The Smoky Mountain Astronomical Society (SMAS) holds star parties at various locations in the area, including Cades Cove, Look Rock, the Tamke-Allan Observatory and Unicoi Crest. Consult the SMAS website for maps and schedules.

> www.blueridgeastronomygroup.com
> www.smokymtnastro.org
> www.visitnc.com/story/adventure-at-north-carolina-state-parks

North Dakota

Much of North Dakota has starry night skies, and, away from the light domes of Bismarck, the southern and central parts of the state offer many good views of the universe. In the northwest of the state, concerns are growing about the impact on the night sky of the oil and gas extraction industry (fracking), especially in the Bakken Shales area from the Canadian border on through Williston and Dickinson. On Interstate 94, which runs right across North Dakota, are Historic Medora and the Theodore Roosevelt National Park, not far from the border with Montana, and the venue for the 3-day Dakota Nights Astronomy Festival. Even here, recent light spill from fracking is seeping over the horizon. See www.npca.org/articles/958-drilling-down for the National Parks' response to the situation.

> medora.com/stay
> www.nps.gov/thro
> www.nps.gov/thro/learn/nature/dakota-nights-festival.htm

Ohio

In 2011, Geauga Observatory Park, near Montville in northwestern Ohio, became an IDA-recognized International Dark-Sky Park. In 2008, the park had taken over responsibility for facilities that had been moved from the growing light pollution of Cleveland to Geauga County in the 1950s. As well as the observatory, there are observing events and astronomy exhibits.

The southern half of the state suffers from stray light from Cincinnati, Dayton, Columbus and other urban areas, but some of the state's parks and forests have reasonable night skies. Try Burr Oak State park near Glouster in southeastern Ohio, Stonelick Lake in the southwest, or the reportedly very dark Wayne National Forest near Graysville in the east. The Observing Sites website below gives further details of these locations. The Cincinnati Astronomical Society based in Cleves near Cincinnati gives details on its website about its observatories, dark sky site and events.

> www.geaugaparkdistrict.org/parks/observatorypark.shtml
> www.observingsites.com/ds_oh.htm
> www.cinastro.org

Oklahoma

In the Oklahoma Panhandle (the long strip of the state projecting from its northwestern corner) is the Black Mesa State Park, one of the darkest areas in Oklahoma and the home of the annual Okie-Tex Star Party, hosted by the Oklahoma City Astronomy Club. The event is held in late September or October, and night skies there, at Camp Billy Joe near Kenton, are Bortle Class 1. The stars are bright from horizon to horizon in this fairly remote place, making it very popular with astrophotographers.

In the southeastern quadrant of Oklahoma is a large area of very dark skies, well away from the lights of urban centers such as McAlester and Fort Smith (Arkansas). A trip along State Highway 144 in Oklahoma's Little Dixie region could make for some fruitful star watching, as for all its length between Honobia (head north from there on Indian Highway) and Talihina it is in a Bortle 1-2 zone. This drive is about an hour away from McAlester, which might be a convenient base for touring this part of the state. The McAlester tourism website has useful information on accommodations. Nearer to the town is the 8000-acre Robbers Cave State Park, whose name recalls a former hideout there of outlaws Jesse James and Belle Starr. The park, in the Sans Bois Mountains, is 5 miles (8 km) north of Wilburton on State Highway 2, and is a recognized dark sky site on www.observingsites.com/ds_ok.htm.

> www.okie-tex.com
> www.cityofmcalester.com/visitors/tourism
> www.stateparks.com/robbers_cave.html

Fig. 3.55 Looking down on Crater Lake, Oregon. (Photo: David Harris)

Oregon

Astronomers speak highly of the relatively remote Crater Lake National Park in the High Cascades of south-central Oregon. The park has local hotel accommodations and food outlets, cabins, and campsites, some of them overlooking the partially flooded 6-mile-wide crater (Fig. 3.55), formed 7000 years ago. The area is snow-bound for much of the year; cross-country skiers might enjoy the challenge. It is many miles from any town. Low air humidity and infrequent cloud cover ensure transparent skies. BAA member David Harris writes: "While waiting for darkness in the Crater Lake National Park I saw some 'clouds' that turned out to be the Milky Way, and that was visible an hour before it got good and dark." On clear, moonless nights, according to the Crater Lake administration, "the stars are too numerous to count. Venus and the Milky Way appear to cast shadows. Views of the night sky from the 7,000-foot elevation at Rim Village are optimum because of the low density of tree cover and the unobstructed view of the horizon in all directions, created by the prehistoric collapse of Mount Mazama."

A useful list of dark sky sites in the state is on the website of the Rose City astronomers, with information about and directions to the Oregon Star Party, which takes place on Fox Reservoir Road in Crook County, at the center of the state, between Mitchell and Paulina. This area is excellently placed for a view of the American total solar eclipse of August 2017.

The Deschutes River State Recreation Area has been labeled "stunning" by visiting astronomers, and an extensive field near campsites T73-T75 and a large parking area are ideal for observing after lights are out.

> www.craterlakeinstitute.com/planning-visit/activities/stargazing.htm
> hotelguides.com/oregon/crater-lake-national-park-or-hotels.html
> www.rosecityastronomers.org/sp/index.htm
> www.oregonstateparks.org/index.cfm (Deschutes)

Pennsylvania

Just to the north of Abbott Township, Pennsylvania, possibly the darkest place in the state, is the 82-acre/33-hectare Cherry Springs State Park. High on the dissected Allegheny Plateau, 2300 feet (700 m) above sea level, the park is located on Pennsylvania Route 44 within the extensive Susquehannock State Forest, a vast expanse of rolling, tree-covered highlands. Cherry Springs was awarded International Dark-Sky Park status by the IDA in 2008. BAA members who have visited this area applaud its wonderful skies: "This has got to be one of the best observing sites in the eastern half of the USA!"

The Cherry Springs Star Party (CSSP), sponsored by the Astronomical Society of Harrisburg, takes place in summer. The Cherry Springs Astronomy Field has low horizons and unobstructed views, is open all year and claims to have "60 to 85 nights each year of ideal conditions for observing." Two major annual star parties draw in large numbers of observers. The site has observatory domes, an amphitheatre, stands for setting up instruments and electricity hook-ups. The 'serious stargazing' web pages give information on overnighting at the park, accommodations and other facilities for visitors. As has happened elsewhere in America (see North Dakota entry), there seems an increasing threat of light pollution on the horizon at Cherry Springs caused by light spill from fracking.

The Black Forest Star Party (BFSP) follows the CSSP in autumn and is sponsored by the Central Pennsylvania Observers. The 2002 BFSP remains a memorable one, since on the Saturday night of the event there was an active, dramatic display of the northern lights.

The area around Warren, in the Allegheny National Forest to the west of Cherry Springs, also has a good report on observingsites.com. Here there are campgrounds and primitive camping sites, and skies have been rated "nearly as good as Cherry Springs."

> www.dcnr.state.pa.us/stateparks/findapark/cherrysprings
> www.dcnr.state.pa.us/stateparks/findapark/cherrysprings/serious-stargazing/index.htm
> www.astrohbg.org/CSSP
> bfsp.org
> www.onlyinyourstate.com/pennsylvania/pa-stargazing

Rhode Island

Rhode Island is the smallest of the U.S. states, and the second most densely populated; it is no surprise that its skies suffer from urban light pollution. The Frosty Drew Observatory at Ninigret Park, Charlestown, is a public-access facility and on Friday nights offers astronomy sessions, both inside and outside the observatory, sometimes until quite late into the night. Its Sky Theatre is used for presentations. Astronomers operate tours of the premises. Visitors are encouraged to bring their own telescopes and binoculars and are expected to share them with others. The observatory's website mentions the likelihood in summer of mosquito nuisance in the area, so dress appropriately and be forearmed with repellent. Already applied is best, as spraying near telescopes is a definite 'no-no'!

Skyscrapers, Inc. (the Amateur Astronomical Society of Rhode Island) operates the Seagrave Memorial Observatory in North Scituate, Rhode Island. On Saturday nights, if clear, the public is invited to observe the night sky there free of charge.

> frostydrew.org/observatory/visiting.php
> www.theskyscrapers.org/free-public-observing-at-seagrave-memorial-observatory

South Carolina

In the extreme northwest of South Carolina in the area around Lake Jocassee and Bad Creek, astronomers sometimes go to find some of the state's darkest sites. Duke Energy, operators of the local hydroelectric project, own much of the land around here, but there is access at certain points, for example by way of the Devil's Fork State Park, to Lake Jocassee. Contact Duke Energy for information about access to roads around the Bad Creek Lake, reputedly the darkest part of the area.

The Santee River area in the southeastern part of the state between Jamestown and the Santee Coastal reserve will be in the direct path of the Moon's shadow in August 2017. This total solar eclipse could be incorporated into a summer stargazing visit.

> southcarolinaparks.com/devilsfork/introduction.aspx
> www.duke-energy.com/south-carolina/service.asp

South Dakota

The Badlands National Park in southwestern South Dakota is a place of almost alien beauty, with deeply incised and eroded pinnacles, buttes and clefts of layered sedimentary rock. There are many places within the park, especially in its southern reaches, well away from urban areas, where the stars are unspoiled. South Dakota Highway 44 between Rapid City and Wanblee passes through this area.

On Friday through Monday nights in summer, the Cedar Pass Campground Amphitheater is the venue for night-sky observing beneath the Milky Way, with telescopes provided and constellation tours. The annual summer Badlands Astronomy Festival is a well-attended event, focusing on, in the words of the National Park Service, "the continued protection and enjoyment of our incredible night skies as a precious natural resource." The festival events do not require advance registration or tickets, but consult the website for latest weather updates and advice.

To explore the area, you might choose to base yourself in Rapid City, 45 miles (72 km) from the park. Stay in the park itself at Badlands Cedar Pass Lodge, one of many sponsors who have made the festival possible, or at the Cedar Pass Campground or Sage Creek Campgrounds.

The last two websites listed below show work by astrophotographers in southern SD.

> www.visitrapidcity.com
> www.nps.gov/badl/planyourvisit/night-sky-program.htm
> www.nps.gov/badl/planyourvisit/lodging.htm
> cbegeman.blogspot.co.uk/2011/06/spring-stargazing.html
> dakotalapse.com

Tennessee

The Pickett-Pogue International Dark-Sky Park (see Chap. 1) in Fentress County, northern Tennessee, became a silver-tier International Dark-Sky Park in 2015. A Clean Tennessee Energy Grant helped to ensure that the local lighting was upgraded to IDA standards. All-year round astronomy takes place in this remarkably dark area, and the main astronomy field is near Pogue Creek Canyon State Natural Area's parking lot on Pickett Park Highway (Fig. 3.56). The field has an electrical outlet for telescopes. There are several different places to stay within the park, which offer 20 cabins from the rustic to the deluxe, and more than 30 campsites with laundry facilities. Nearly all have electrical hook-ups. For very useful information on finding the main observing site and other nearby dark sky observing locations see celestial-imaging.com/archives/2356.

The Cumberland Astronomical Society, which meets at Volunteer State Community College in Gallatin, Tennessee, has information on its website about various astronomical events, including the Tennessee Spring Star party, at Fall Creek Falls State Park. The park is near Spencer, about 70 road-miles north of Chattanooga.

The Barnard-Seyfert Astronomical Society (BSAS) of Nashville holds public star watches at out-of-town locations. The society's name recalls Edward Emerson Barnard (1857–1923), a prolific American comet discoverer born in the city, and Carl Keenan Seyfert (1911–1960), investigator of active galaxies, who died there.

Fig. 3.56 Pickett-Pogue by day. (Photo courtesy Pickett-Pogue International Dark Sky Park)

darksky.org/idsp/parks/pickettpogue
www.cumberlandastronomicalsociety.org/FallCreekFalls/FallCreekFalls.
   html
www.tnstateparks.com/FallCreekFalls
www.bsasnashville.com

Texas

Big Bend National Park is one of America's largest, and lies far to the south at the
U.S.-Mexican border on the Rio Grande (Mexican, Río Bravo del Norte). Without
doubt, its remoteness and determined anti-light-pollution strategies make it, in the
words of its website, "one of the best examples of primeval night skies available
today" in the lower 48 states. A very low population density and wild terrain bring
the kind of night skies that qualified it in 2010 as an IDA gold-tier International Dark-
Sky Park. Frequent stargazing events occur here throughout the year, in association
with local astronomy groups and the McDonald Observatory near Fort Davis to the
north. The Texas Dark-Sky Festival is held at Fort Leaton State Historic Site,
Presidio, to the west of the park. Details of camping and accommodation are at visit-
bigbend.com/vcategory/big-bend-national-park-lodging.

Fig. 3.57 The Coathanger asterism in Vulpecula, photographed in dark skies over Yancey, southern Texas. (Photo: Carey Johnson)

The Texas Night Sky Festival is held at the state's first IDA Dark-Sky Community, Dripping Springs, west of Austin (Fig. 3.57).

About 15 miles (25 km) north of Fredericksburg is the Enchanted Rock State Natural Area, another Texan International Dark-Sky Park. Enchanted Rock has hosted star parties for park visitors since 2011. The area has some fascinating geology, and its vast dome of pink granite is a magnet for hikers and climbers. At night, its rural dark skies attract stargazers from all around.

> tpwd.texas.gov/spdest/programs/dark_skies
> www.fortdavis.com/guesthouse.html
> presidiotx.us/ufo/?page_id=969
> darksky.org/idsp/communities/drippingsprings
> tpwd.texas.gov/state-parks/enchanted-rock/dark-skies

Utah

About 37 miles (60 km) to the northwest of the dark sky haven of the Hovenweep National Monument (see Colorado entry), there is another place of striking landforms and beautiful starry vistas—Natural Bridges National Monument. In March 2007, this site became the first International Dark-Sky Park. The darkness of its

Fig. 3.58 A stunning image of the Milky Way over Natural Bridges National Monument. (Photo: Jacob W. Frank)

nightscapes can be seen in Jacob W. Frank's wonderful photo (Fig. 3.58), depicting what its website calls "the stunning river of light formed by the Milky Way rising over Owachomo Bridge." Basing yourself in Blanding, Utah, to the east of the monument, visit its three surreal bridges of Permian sandstone, undercut by flooding streams and a reminder of the power of erosion; then wait for nightfall and, if the sky is clear, savor those stars.

Utah also has the world's first IDA bronze-tier International Dark-Sky Park. In the Wasatch Range north of Ogden is Weber County North Fork Park, well known for its high-altitude ski trails. It has six large accommodation areas and large numbers of individual campsites, beneath some of the best night skies in northern Utah (Fig. 3.59).

In the south of the state, the relatively remote and elevated Bryce Canyon National Park takes pride in what it calls its "viewshed," the vast area visible from the park. The night here is "as dark as dark can be – on a clear day at Bryce you can see nearly 200 miles to the Black Mesas in eastern Arizona. On a clear dark night, you can see 2.2 million light years, or 527,000,000,000,000,000 miles, to the Andromeda Galaxy."

> www.nps.gov/nabr/learn/nature/darkskypark.htm (Natural Bridges) www. blandingutah.org
>
> www.co.weber.ut.us/parks/nfpark.php (North Fork Park)
>
> www.nps.gov/brca/learn/nature/lightscape.htm (Bryce Canyon)

Fig. 3.59 Scorpius with its red giant star Antares over Utah. (Photo: Nick Hart)

Vermont

The annual Stellafane Convention in Vermont has become synonymous with dark sky gatherings, and it is, presumably, the only stargazing event to have an asteroid (number 3140) named for it! In the early 1920s the Springfield Telescope Makers (STM) club of Springfield in southwestern Vermont built a clubhouse they named Stellafane (Latin for "shrine to the stars"), and this has now come to refer to the club's land and buildings on Breezy Hill, to the southeast of the town. The convention is usually held there during the weekend of the new Moon closest to the maximum of the Perseid meteors in August; it brings together very large numbers of amateur telescope makers and astronomers. Most activities are based at Stellafane East with, among other facilities, the McGregor Observatory. Ken Slater, STM vice-president, says "check our website's events calendar for current information. Public events are highlighted in yellow. Also, Saint-Gaudens National Historic Site in New Hampshire hosts a free annual Star Party in partnership with the Springfield Telescope Makers" (see New Hampshire entry).

stellafane.org/news/index.html#events
www.nps.gov/saga/learn/news/star-party.htm

Virginia

At the center of southern Virginia is the Staunton River State Park, an expanse of grassland and forests about 25 miles (40 km) north of the border with North Carolina. The park is popular with hikers and other outdoor types, but its campgrounds and historic cabins now also welcome astronomers. On clear nights, the area's excellent skies are well worth the trip. The park's first Staunton River Star Party was held in 2011, and growing numbers of enthusiasts now attend this annual event.

Richmond Astronomical Society's website mentions dark sky sites in Virginia, as does the Blue Ridge outdoors site. Several sources suggest that the best dark sky in the state may well be in and around Groundhog Mountain Park in southern Virginia, on the 469-mile Blue Ridge Parkway.

> www.chaosastro.com/starparty (Staunton River Star Party)
> richastro.org
> www.blueridgeoutdoors.com/go-outside/dark-sky-destinations
> www.virginia.org/Listings/GroundhogMountain

Washington

The Washington Trails Association has a useful web page detailing dark sky places in the state. One of the best is arguably Lake Crescent in the Olympic National Park, on the Olympic peninsula. The area has many campgrounds, cabins and hiking trails. Informal and organized star watching go on in the park.

Near the small town of Goldendale, in the far south of Washington, an observatory stands on a hill in what the IDA terms "a spot of darkness gazing at the Milky Way, and the primary night-sky interpretive site in the Washington State Parks System." This facility is situated in the Goldendale Observatory State Park, two hours' drive west from Portland, Oregon, and is visited by thousands of astronomy enthusiasts every year. The observatory area became an IDA silver-tier Dark-Sky Park in 2010. As well as the observatory with its large 24.5-inch reflecting telescope in the South Dome, there is a picnic area, an education center and an amphitheater. Smaller on-site telescopes include portable Dobsonians and a solar telescope.

The Deschutes River State Recreation Area (see Oregon entry) with its campsites is not far from Goldendale.

> www.wta.org/go-hiking/seasonal-hikes/summer-destinations/wtas-dark-
>   places-digest
> www.nps.gov/olym/planyourvisit/visiting-lake-crescent.htm
> www.nps.gov/olym/planyourvisit/nightsky.htm (Olympic National Park)
> www.goldendaleobservatory.com

West Virginia

At 4863 feet, Spruce Mountain, within the Spruce Knob-Seneca Rocks National Recreation Area, is West Virginia's loftiest peak. Spruce Knob, at its summit, has a definite Alpine feel, with spruce forests, strewn boulders, meadowland and the Observation Tower where visitors admire both sunsets and the night sky. This ranks high on the list of darkest sites east of the Mississippi.

This area of the Southern Appalachians is also the site of the yearly Almost Heaven Star Party (AHSP), held at the Mountain Institute near Spruce Knob. The party offers trade stands, talks by experts, off-site activities and of course stargazing. Not far away to the south is the Green Bank National Radio Astronomy Observatory (NRAO), in the officially designated National Radio Quiet Zone, an area of the country where radio transmissions are restricted to aid scientific research. Bookable in advance on the GBT website are a bus tour of the Green Bank facilities, including its giant radio telescope, hands-on activities and StarLab inflatable planetarium sessions. Admire the constellations by night, too, at the annual NRAO/Central Appalachian Astronomy Club/Kanawha Valley Astronomical Society Green Bank Star Quest, a combined optical and radio star party. Associated campsites are close to the NRAO.

> gotowv.com/dark-sky-wv
> www.ahsp.org
> public.nrao.edu/tours/visitgbt
> www.greenbankstarquest.org

Wisconsin

The northern strip of Wisconsin and the central part of its western border with Iowa are the areas where stars shine brightest. The Chequamegon-Nicolet National Forest (CNNF) area in northeastern Wisconsin is known for its night skies. Try the area around Phelps by North Twin Lake on Highway 17. The CNNF website gives details of camping and accommodations.

In the west, Rush Creek State Natural Area, near Ferryville, is a swathe of dry prairies and woodland by the steep bluffs overlooking the Mississippi River. Light-pollution maps show this to be one of western Wisconsin's best stargazing areas. Camping is possible here, and Ferryville and Prairie du Chien, further down the river, have hotels. Although night skies towards Madison and Milwaukee in the east have fewer stars, visitors to the state might enjoy visiting one of America's sites of special astronomical interest, Yerkes Observatory, a facility of the University of Chicago's Department of Astronomy and Astrophysics. Founded in 1897, Yerkes has the world's largest (40-inch/100-cm) refracting telescope, and famous astronomers who worked here include Edward Barnard, Edwin Hubble and Carl Sagan. See the website for information about opening times, tours and events.

wisconsinastronomy.org/clubs.html
www.fs.usda.gov/cnnf
www.wisconsinbirds.org/trail/sites/rushcreek.htm
astro.uchicago.edu/yerkes

Wyoming

Astrophotographer Bob Neville speaks highly of the majestic skyscapes of Yellowstone National Park (Fig. 3.60) in Wyoming: "The hotel and lodges at Yellowstone Lake were ideal for observing the night sky. The park is not developed except for a few hotels. There are no streetlights for many miles and, at an altitude of 7,750 feet, the sky is virtually black."

Nearly all of Yellowstone National Park is in Wyoming, the rest being in Montana and Idaho. The 3468-square-mile park is larger than Rhode Island or Delaware. Its highest summit is Mount Washburn (10,243 feet/3122 m). Remember that even in early autumn here the nights get cold. Snow is likely in autumn and

Fig. 3.60 Perseus and the Pleiades from Yellowstone. (Photo: Bob Neville)

inevitable—usually in great quantities—in the winter, with limited access and transit.

Stargazing spots recommended by astro-tourists and the park itself include Mount Washburn, Upper Geyser Basin, Old Faithful Visitor Education Center, Mammoth Hot Springs and the Madison Amphitheater. The park runs summer astronomy programs, and telescopes are available.

Not far to the south of Yellowstone Park is the mountainous Grand Teton National Park, which also has good dark areas at night. This area is known to many for the great daylight fireball (sometimes known as the Grand Teton Meteor: see the video at asv.org.au/meteor) that raced across its sky on August 10, 1972. The American total solar eclipse will be seen, weather permitting, from the park in August 2017, with two minutes and twenty seconds of totality (the Sun completely covered by the Moon). Plan carefully if you intend to observe this from Grand Teton—you won't be alone!

Visitors to both Grand Teton and Yellowstone National Parks often arrive at Jackson Hole Airport, the largest and busiest commercial airport in Wyoming. Jackson Hole is a wide valley through which the Snake River winds. Away from the town of Jackson itself, the area shares the wonderful dark skies of the Grand Teton Park.

In central Wyoming, west of Casper, scenic U.S. Route 287 passes through a swathe of night-time darkness, with Jeffrey City somewhere near its center.

> www.yellowstonepark.com/2014/01/gaze-stars-night-skies
> www.nps.gov/yell/planyourvisit/calendar.htm
> www.nps.gov/grte

Having explored the UK and the United States for sites where the stars shine, the intrepid astro-tourist might wish to venture further afield and seek out some other places, both traditional and more recently developed, where the night skies remain unspoiled and the stars and the aurora may be seen. Here is a selection of some recommended dark sites in European countries outside the UK.

Austria

This author attended the Eighth European Dark Sky Symposium at Vienna's Kuffner Observatory in August 2008, coincidentally at a time when, in a city-wide relighting scheme, new flat-glass lamps were replacing the thousands of distinctive fluorescent tubes hanging from cables above the city's streets. Emissions from these old lamps were poorly directed, and the mayor of the city gave a press interview at the symposium welcoming the new technology.

The Kuffner Observatory, a venerable building completed in 1885 and once one of the principal astronomical facilities of the old Austro-Hungarian empire, saw its role gradually altered in the late nineteenth century from observational to educational, as the cosmopolitan city grew around it and public lighting proliferated. High on the 450-m Gallitzinberg, a wooded hill in the western suburb of Ottakring, the observatory retains its large telescopes and transit instruments, kept in pristine condition for inspection by the many visitors that pass through its doors.

Not far from the capital is a noteworthy project—the Austrian Dark Sky Oasis at Großmugl (Sternlicht Oase Großmugl), with its huge ancient tumulus, a raised mound of earth and stones over a burial site, called the Leeberg (Fig. 3.61), dating from 2500 years ago. Here there is a star walk (Sternenweg) designed by Project Nightflight and built in collaboration with the municipality.

Bylaws have been enacted by the local council in Großmugl to protect the night sky in the vicinity, and its outdoor lighting is strictly controlled. The delegates at the symposium were bussed out to the site in the evening and, by the light of mobile phones, picked their way along pitch-dark trackways to the Leeberg to see the few stars available to them as the clouds had rolled in, almost, it seemed, on cue.

Fig. 3.61 Comet Lovejoy (the nebulous object at center right) over the Leeberg. (Photo: Guenther Wuchterl, Kuffner-Sternwarte.at)

The Leeberg would certainly be a site worth visiting for skywatching on cloud-free nights, but remember to take a red light for the walk up to the tumulus (red, of course, to preserve night vision).

Further afield in Austria are the mountainous Gesäuse and Kalkalpen national parks, in the states of Styria and Upper Austria, respectively. Both are within the East Alpine Starlight Reserve. The Gesäuse National Park is famous for its white water expeditions on the great Alpine River of the Enns, which descends more than 150 m along the 16 km of its steep-walled gorge. Its few small communities contribute little light pollution. This means that the sky quality in the Starlight Reserve area is among the world's best, and SQM readings of up to 21.85 (borderline Bortle Classes 1 and 2) have been recorded.

The Kalkalpen (Limestone Alps) National Park, harbouring central Europe's largest forested area, has typical karst scenery with many caves, springs, gorges and incised mountain streams. Having admired its orchids and chamois deer by day, and if you are fortunate enough to have clear skies you might well see the zodiacal light from open areas just before dawn or after dusk, and admire the park's Milky Way vistas in skies that have been graded as "possibly the best in Central Europe" by Austrian dark sky activists.

> www.clickstay.com/blog/travel-to-the-stars
> www.project-nightflight.net/the_grossmugl_star_walk.pdf

Belgium

In the predominantly French-speaking Walloon region of Belgium is an area of German-language speakers, to the east near the border with Germany. Its main town is Liège (Lüttich), and an hour's drive to the southeast of the city brings you to Bütgenbach, near Malmédy and close to the frontier. Belgian dark sky observers say that this is probably the darkest area in the country, many parts of which remain light-polluted. Plans to incorporate the Bütgenbach area into a larger dark sky enclave of some kind have been discussed. Bütgenbach has hotels, and there are several campsites nearby, some near its artificial lake.

In Belgium's Flemish region is an island of relative darkness: the Blankaart Nature Reserve in West Flanders about 25 minutes' drive northwards from Ypres (Ieper). Located southeast of Diksmuide, in the village of Woumen, the Blankaart Reserve is on the right bank of the River Yser. It is among the oldest protected natural sites in Belgium, and a stopover for countless migratory waterfowl. Its fenland terrain is not conducive to rambling, and much of its area can flood over in the winter, but find some firm ground and clear horizons, and you will see the stars, weather permitting.

As is happening in the UK, more and more regions in Belgium are regaining their stars late at night as municipalities dim or switch off lighting to save energy and money. That the authorities will consider doing this, and that residents on the whole accept it, says much for the raising of awareness among the general public of the value of the night; the work of the very active Belgian Dark Skies movement over the last thirty years has helped to change public attitudes.

> en.wikipedia.org/wiki/Bütgenbach
> en.camping.info/belgium/campsites
> www.routeyou.com/en-be/location/view/47570476 (Blankaart)

Bulgaria

Away from Sofia and Plovdiv, much of Bulgaria offers rural dark skies. In the southwestern corner of the country is the Bulgaria National Astronomical Observatory (BNAO) at Rozhen, which is about 200 km south of Sofia. This facility has been conducting serious astronomical research since March 1981, and it prides itself on being the biggest astronomical facility on the Balkan Peninsula, and even southeastern Europe. Its main telescope is a 2-m reflector, and there are two other large optical telescopes with which studies of both the night sky and the Sun are carried out. Meetings, school visits and workshops are regular features of the BNAO program, and its visitor center has a lecture hall, a dedicated astronomical museum and a 30-cm demonstration telescope. The observatory's website mentions that "in the 2-m telescope dome, lectures, talks and demonstrations with the telescope are carried out for individual tourists and organized groups."

Enthusiasts seeking a long-term commitment in astronomy and a change of scenery may be interested to know that a dark sky village has been proposed for central Bulgaria, in the Stara Planina Mountains, reputedly one of the darkest areas in Europe. The plan is to build about 50 environmentally engineered 'observatory houses,' with the needs of astronomers and sky watchers in mind.

> www.nao-rozhen.org/visit/fr4_en.html
> www.skyvillagebg.com

Croatia

The coastline of Croatia, and especially the many elongated islands running parallel to it, offers some good dark skies, but one of the southernmost islands off the coast near Dubrovnik is famed among European amateur astronomers for its night-time vistas. This is Lastovo, called by its tourist board the "Island of Bright Stars." In September 2006 the Croatian government created a new zone of protection for the natural resources of this area, designating the island of Lastovo and its archipelago of smaller islands as the Lastovsko Otocje Croatian Nature Park.

The municipality replaced old and wasteful street lighting with new and less bright lamp types, which considerably improved the visibility of the stars for

Fig. 3.62 Lastovo Milky Way. (Photo: Andrej Mohar, Herman Mikuž, Dark Sky Slovenia)

observers in its vicinity. Now, Lastovo is reputed to have the darkest skies in Croatia and, according to more than one source, the darkest in Europe. Taken with a camera from Lastovo's highest point, which is 400 m (1300 feet) above sea level, the illustration (Fig. 3.62) shows the Milky Way reflected in the sea to the south of the island. To underline the fact that stray light can travel a long way, there is a faint trace of it in the photo—its source the Italian city of Bari, in Italy, which is 180 km (112 miles) across the Adriatic Sea.

> www.observatorij.org/dogodki/Lastovo2010
> www.lastovotravel.com/accommodations-lastovo

Czech Republic

The first two-nation Dark Sky Park in the world lies in the upper parts of the Izera and Jizerka valleys of the Czech-Polish borderland. The Izera Dark Sky Park was established in 2009 by a collaboration of several Czech and Polish organizations, including the Czech Republic's Astronomical Institute of the Academy of Sciences, the Jizerské Hory Protected Landscape Area, the Astronomical Institute of the University of Wrocław and the Polish and Czech State Forestry Commissions. The very thin population density of the region and low level of outdoor lighting within the valleys, together with the fact that the local mountain ridges act as natural shields against stray light from more distant towns and villages, make this a dark place, and the park has the added benefit of being easily accessible for tourists wanting to explore this interesting and environmentally unique region.

Extensive peat bogs and forests are the main features of the landscape, and the dark sky zone was created with the specific intention of promoting conservation of its natural heritage while raising public awareness of the value of the night sky. Places where tourists can stay include Orle, Chatka Górzystów and Jizerka, and in all of these locations can be found fliers about the Izera Dark Sky Project, information about organized stargazing and the night sky brightness monitoring that goes on here using all-sky cameras. The Czech Republic is becoming an important state within the ranks of those that are taking active measures to tackle light pollution. In 2013 the Beskydy Dark Sky Park at the republic's eastern tip came into being, followed in 2014 by the Manetin Dark Sky Park to the southeast of Karlovy Vary (Carlsbad) in the western Czech Republic. In the Beskydy Park it is estimated that up to 2000 stars crowd the night skies from certain locations, and in the Manetin area SQM readings at more than 200 sites have been taken that confirm the unusual depth of darkness of the sky over most of the park.

> www.izera-darksky.eu
> www.czechtourism.com/c/manetin
> www.boto.cz (Beskydy—open page via English translator)

The darkest parts of the island of Cyprus are at its easternmost and westernmost tips. Akamas National Park is located at the western end of Cyprus, on the hook-like promontory of the Akamas Peninsula, famed for the Baths of Aphrodite on its rugged eastern edge. Rumors are that plans were afoot to set up a dark sky zone within the park, but there is no confirmation of this currently. However, searches of newspaper articles and tourist reviews indicate that this is still, in the words of a *Daily Telegraph* piece, a place of "pitch black skies at night." The Cyprus Astronomical Society is a good source of advice for stargazing possibilities on the island, and another good way to assess these is to look at www.lightpollutionmap.info.

Since 2008, the Kition Planetarium and Observatory near Larnaka in southeastern Cyprus has been, to quote its publicity, "encouraging awareness of astronomical issues among the general public, as well as to get people involved in the world of discovery, exploration and wonder." The planetarium has a 5-m dome, with an adjacent observatory, and there is also a weather station and conference room, as well as an exhibition of fossils and geology. Visits are by appointment, or through tour operator groups. Events may be held in English, Russian and Greek at various times of the month.

> www.paphosfinder.com/cyprus-info/akamas.htm
> larnakaregion.com/page/planetarium
> www.CyprusAstronomy.com

Denmark

In 2016, Denmark, a rather light-polluted country, has a serious candidate for membership of the IDA dark sky club. It seems that the island of Møn, south of Copenhagen, is aiming for recognized Dark Sky status, possibly as a Dark Sky Park or Community. The light pollution maps suggest that Møn is reasonably starry at night, and may have some of the best views of the night sky in the country. A local astronomy group, Dark Sky Møn (DSM), is working towards preserving its area's night sky for both present and future observers.

Tom Axelsen is head of the South Zealand Astronomy Association and an administrative member of DSM. On the association's website he states that "the quality of the night skies above Møn and Nyord is extremely good by Danish standards. The Milky Way, for instance, is so huge and clear that it surprises even experienced amateur astronomers." Accommodation is available at the local Klint Resort.

> www.darksky-moen.dk (may require translation tool)
> www.moensklintresort.dk

Estonia

At first glance Estonia appears, on the satellite maps, to be the most light-polluted of the three Baltic States, but closer inspection reveals that the large Baltic Sea island of Saaremaa (Osel), 2673 square km (1032 square miles) in area, has plenty of remaining dark areas. Saaremaa is in fact the largest island in Estonia. The small capital of the island is Kuressaare, with an airport handling regular flights to and from the country's capital, Tallinn.

What makes Saaremaa doubly interesting for astro-tourists is that, having seen the stars from rural parts of the island, they may visit during the day a grouping of nine meteorite craters near the village of Kaali. The largest of the craters, Kaali Järv (Lake Kaali), is 110 m (360 feet) in diameter.

Eight smaller craters, all within 1 km of the main crater, are also associated with this impact event. They range in diameter from 12 to 40 m and vary in depth from 1 to 4 m. The object that formed these craters came in from the northeast, broke up in flight in the atmosphere and scattered fragments across the area, leveling forests within a radius of 6 km (3.7 miles). Analysis of silicate spherules found in the boggy terrain of the region suggests that the impact probably occurred about 7500 years ago. The total mass of the incoming meteoritic body has been estimated at between 20 and 80 million tons. It has been calculated that the energy of the impact could have been equivalent to that released by 20 kt of TNT, comparable to the Hiroshima atomic explosion.

Saaremaa is linked by ferry to Virtsu on the Estonian mainland. In some winters it is possible to drive by road to Saaremaa from the mainland, and there are regular bus services from Tallinn, Pärnu and Tartu. Plans for a Saaremaa Bridge or Saaremaa Tunnel connecting with the mainland have been proposed.

www.visitestonia.com/en/kaali-field-of-meteorite-craters

Finland

Skies over Finland's capital city Helsinki are, say local astronomers, somewhat light-polluted. Indeed, the whole southern half of the country, where most of its towns are, can promise little in the way of pristine night skies, according to light pollution maps such as darksitefinder.com/maps/world. Further north, when winter has guaranteed long hours of darkness at night in this Land of the (summer) Midnight Sun, it's a very different picture. The lakes (there are nearly 200,000 of them) and the stars are in good supply in northern Finland. In northeastern Finland, abutting the border with Russia—nearer to Murmansk than to Helsinki— is the beautiful Arctic fell country of the huge and remote Urho Kekkonen National Park, on an ancient Sami route to the Barents Sea. Whether it's the Midnight Sun you prefer or the wonders of the aurora, holiday companies offer trekking, cross-country skiing, aurora watching from glass igloos, pinewood cabins and more. To find out more, search "Finland Aurora Tourism" and "Finland Dark Skies" on the Internet.

France

In December 2013 the world's second largest International Dark Sky Reserve (Réserve Internationale de Ciel Étoilé du Pic du Midi) came into being in the Pyrénées Mountains. Confirming the reserve's status, the IDA applauded the sky protection strategies of the Dark Sky team at the Pic du Midi (2877 m/9439 feet), upon which sit some of France's largest telescopes. These include the famous Lyot coronagraph, a pioneer in solar coronal studies, and several big reflectors in dedicated domes that can be seen on a tour of the facility. Inching around the domes' exteriors on metal mesh walkways with vertiginous rock faces below is not for the faint-hearted!

The area of the Pic du Midi lies within a UNESCO World Heritage Site and the Pyrénées National Park. The daytime views in all directions of the Pyrénées from the observatories' viewing terraces are stunning, and night-time observing takes place there by arrangement (www.picdumidi.com/soirees-etoilees: translation tool available). Local French contacts insist that the night skies over this region of the Central Pyrénées are "superb… Milky Way visible soon after sunset… worth a trip from anywhere." The area immediately around the foot of the mountain is built up, as any tourist destination tends to be, but further afield (and in almost any other part of this mountain range) there are wonderful starscapes to be found. For example, near Llauro not far from Perpignan in the Catalan country of the eastern Pyrénées, the stars of Scorpius and Sagittarius can be seen rising, higher up than they could ever be in southern England, shimmering above a narrow, pitch dark road, the D615. The many glorious star clusters and nebulae in these constellations are easily visible to the unaided eye as they clear the dark silhouettes of the distant mountains. You can base yourself at the welcoming Hotel Néoulous in Le Boulou. For the central Pyrénées area you might consider accommodation in Bagnères-de-Bigorre, and for the western part, Oloron-Sainte-Marie.

French amateur astronomers cherish the so-called "Quercy Dark Triangle" in the Lot *département* of south-central France, an island of darkness on the light pollution map, surrounded by brighter patches. Here, public stargazing is encouraged, and local villages are evolving anti-light-pollution codes of practice and declaring themselves to be "Villages of the Stars"; examples are Carlucet, Reilhac, Vidaillac and Beauregard. Not far away, in the Dordogne *département* near the *bastide* village of Monpazier in the Pays de Bergerac, may be the darkest sky you will ever see in mainland Europe. Starclouds and dust lanes of the Milky Way can be visible in great detail to the unaided eye. Local Neolithic cave dwellers must have wondered at this, tens of thousands of years before, their only companions stock-still geckos seen dimly by starlight upon a rock, and luminous green constellations in the grass—glow worms mirroring the brightest stars in the sky—with not a human light to be seen anywhere.

> www.tourism-midi-pyrenees.co.uk
> www.france-voyage.com/cities-towns/bagneres-de-bigorre-25726.htm
> www.france-voyage.com/cities-towns/oloron-sainte-marie-25533.htm
> beusse81.free.fr/lileauxetoiles (French site for map showing Dark Triangle
>   of Le Quercy)

Germany

Among other recognized dark sky places in Germany are two silver-tier International Dark Sky Reserves: the Westhavelland in Brandenburg, west of Berlin; and the Rhön, northeast of Frankfurt. The Eifel National Park (North Rhine-Westphalia) is the most recent addition to the IDA list, having received Dark Sky Park status in 2014.

Germany has plenty of other places on its heaths and in its mountains and woodlands where the stars may be enjoyed, for example the Müritz National Park (Fig. 3.63) in Mecklenburg-Vorpommern, between Berlin and Rostock.

South of Brunswick (Braunschweig) in central Germany are the Harz Mountains, an area far less light polluted than many of the surrounding districts. Around Sankt Andreasberg, with its new observatory, skies are attractively dark, and on clear nights there is plenty to see, either with instruments or the unaided eye. In 2011 the

Fig. 3.63 Moonrise over the Müritz National Park. (Photo: Alejandro Sanchez de Miguel)

UNESCO Star Park Program listed Sankt Andreasberg as a recommended observing site, and in 2012, the Dark Sky Group of the Vereinigung der Sternenfreunde (Star Friends Association), the biggest association of amateur astronomers in Germany with more than 4000 members, included the Sankt Andreasberg region in its list of Germany's "top seven observation sites of the highest quality." In 2011, a team under the direction of the Dark Sky Group's director Dr. Andreas Hänel, a leading figure in the activities of the IDA Europe Section, had measured night sky darkness at a location between Braunlage and Sankt Andreasberg. They registered an encouraging SQM measurement of 21.75 (Bortle Scale 3), an excellent reading from the middle of Europe's most populous country (Germany has more than 80 million people). A group of enthusiasts in the Harz region is aiming to have the Harz National Park declared an IDA Dark Sky Park.

Tribute must be paid to such groups and to the members of the IDA Europe Section in various countries, whose combined efforts have contributed to the protection of threatened dark skies on a crowded continent.

> darksky.org/idsp/reserves/westhavelland
> darksky.org/idsp/reserves/rhon
> www.nationalpark-eifel.de/go/eifel/english.htm
> www.germany.travel/en/national-parks/mueritz-national-park.htm
> en.wikipedia.org/wiki/Sankt_Andreasberg_Observatory

Greece/Crete

Sizable tracts of Greece away from Athens still enjoy very dark skies, including the Pindus National Park in the north and the central mountain chain that extends southwards from it. However, and somewhat surprisingly, much of the rest of rural Greece has less than perfect skies. Most reports by returning tourists telling of star-strewn Greek skies dwell on the celestial wonders that can be seen from the Greek islands. Those which have not yet been too 'touristified,' and there are many, retain their stars. On Antiparos, for example, a local astronomer hosts gatherings in his observatory, and on Folegandros a local holiday villa lends a large telescope to guests.

Crete has a long chain of dark places along its southern edge. In the southeast is Agios Stefanos, where, at Sasteria, astro-tourists can use telescopes at a dedicated astronomy site.

> www.tripadvisor.com/antiparos/Telescope/Observatoryfolegandros-villas.
> com/en/stargazing-sea-observation
> www.sasteria.com

Hungary

In the northern Zselic region of southwestern Hungary, the Zselic Starry Sky Park, established in 1976, describes itself as "one of the best places with close to unaltered dark night skies in Hungary." The Duna-Dráva National Park Directorate and the Hungarian Astronomical Association did tremendous work across the Zselic National Landscape Protection Area, collaborating with the local municipalities, hotels and the Lighting Society of Hungary to minimize stray light impact on the park.

Seeing the faint, extended Triangulum Galaxy (M33) with the unaided eye is a huge challenge. Although it appears bigger than the full Moon in the autumn/fall and winter skies, discerning it without optical aid is all but impossible if the sky has any amount of moonlight or artificial light present. At Zselic, M33 can be seen by normally sighted observers, and certainly by using the averted vision technique, on clear, transparent nights when the air is still, and the zodiacal light is often visible in spring and autumn. This is truly a Bortle Class 1 area. Sky darkness measure-ments in the Zselic region are among the best in Europe. Understandably, the Zselic Starry Sky Park became an IDA Dark Sky Park in 2009.

Hungary's first national park, the Hortobágy in eastern Hungary, has Europe's largest continuous expanse of native grassland; it is a UNESCO World Heritage Site and Biosphere Reserve, and a Ramsar Convention area. It is also one of the darkest spaces in Hungary. In 2011 the park, an extensive area of marsh and grassland, was awarded the silver tier Dark Sky designation by the IDA. It has instituted a special monitoring program for its very diverse nocturnal species, and allied to this is a park-wide lighting management plan focusing on the protection and promotion of bird and bat species. For astro-tourists, Hortobágy offers themed night walks, light pollution education sessions and an observatory.

darksky.org/idsp/parks/zselic
darksky.org/idsp/parks/hortobagy

Iceland

Generously illuminated by relatively low-cost electricity from geothermal sources, Reykjavik, the Icelandic capital in the west of the great volcanic island, is not the best place to view the constellations or aurorae. Astronomers returning from aurora hunting trips and from viewing the solar eclipse of March 2015 have commented on Reykjavik's degraded night skies, a situation aggravated when the 'Peace Tower' commemorating John Lennon, on Videy Island north of the capital, is activated, shining pillars of blue light upwards. Most of the rest of Iceland's bare and rugged terrain, not far below the Arctic Circle, has great views of the stars and relatively frequent auroral displays (Fig. 3.64). To see the aurora at its best, try the view from the Thingvellir National Park, an hour's drive away from the capital's lights

Fig. 3.64 An Icelandic aurora. The green hue is due to the presence of ionized oxygen in the atmosphere. (photo: David Strange)

along Routes 1 (North) and 36. An hour and a half along Route 1 (East) is the Hotel Rangá, not far beyond Hella. There is probably not another hotel in the world that possesses its country's largest telescope (in this case an 18-inch reflector).

> www.thingvellir.is/english.aspx
> www.hotelranga.is/stargazing

Italy

Light pollution maps (for example darksitefinder.com/maps/world.html) show that light spills from Milan, Turin and other big cities have compromised the night sky across much of the northern swathe of Italy. However, in the south, especially around one of Italy's oldest national parks, the Sila National Park in Calabria, dark skies can still be found. The Apennine area of small towns and villages to the south-east of Potenza (Basilicata region) is also promising.

The maps also show a long band of comparative darkness in Tuscany. The north of this band is centered on Musci. Its southern part lies around Campigliola. Where is the darkest place in Italy? According to highly respected Italian dark sky campaigner Fabio Falchi, of the Light Pollution Science and Technology Institute (ISTIL), the best night skies are in Sardinia and, on the mainland, in the very north-ernmost part of the country. Here, well away from the big cities, at the border with Austria, is the German-speaking region of Alto Adige-South Tyrol, and to the south and in the foothills of the great mountain ranges, undeveloped places retain their darkness.

Regione Veneto was the first Italian region to enforce a law against light pollution, in 1997. At first, the law had room for improvement, in that its clauses allowed luminaires (as lighting professionals call lamps) to allow up to 3 % of their emissions to escape as "direct upward-directed flux." Even at such low angles, stray light will find its way inexorably into the sky and spoil views of the heavens far away from the light source. More than ten years later, in 2009 (the International Year of Astronomy), the Veneto Region substantially upgraded its lighting law and rewrote parts of it, recognizing the importance of zero upward-directed flux. As Fabio Falchi then wrote: "Galileo used his telescope for the first time in Veneto, and now Veneto has made a huge step forward in protecting the night sky and the night-time environment."

Other Italian regions now have similar ordinances, including the most populated one, Lombardy.

> www.turismo.intoscana.it/allthingstuscany/aroundtuscany/want-to-see-the-stars-come-to-southern-tuscany
> www.inquinamentoluminoso.it (ISTIL; in English)

Latvia

There are four national parks in Latvia. The largest and oldest is the Gauja National Park, famed for its cave systems and grottoes. It is about 50 km (31 miles) to the east of the capital, Riga. Skies here are moderately dark, and the further you can get from Sigulda to the south of the park and Cesis to its east, the better they will be. The marshlands and warm springs of the Kemeri National Park are its main attractions, but starry views do not get much of a mention in its literature. Skies here will be too tainted in the east because of the proximity of Riga and satellite towns.

The Razna National Park, originally set up to safeguard the great diversity of species and habitats of the Lake Razna area, lies to the south of Rezekne, and is far enough away from the built area to be a much better bet. Skies here have little of the distant urban light signatures of Gauja and Kemeri.

However, the winner in the dark-sky stakes in Latvia has to be the coastal Slītere National Park on the Gulf of Riga. The park is 65 km (40 miles) northeast of Ventspils on the large promontory that forms the northeastern part of Latvia. The Slītere National Park is famous for its so-called Blue Hills (Zilie Kalni), dune fields and coniferous and broadleaf forests, harboring many rare species of plant and animal. It is one of the last refuges of the Eurasian lynx and the ringed seal. You might see a little light leaking from towns in Estonia and northern Latvia far away across the gulf, but Slītere's has to be the best accessible night sky in any of the country's national parks.

www.latvia.travel/en/article/national-parks-latvia

Lithuania

Eržvilkas, in Taurage County, is a small town of about 600 people, about halfway between the major towns of Kaunas and Klaipėda. It lies at the western end of a strip of natural darkness in west central Lithuania, and although there is apparently no dedicated, organized dark sky facility here, this is probably the best area in the country from which to observe the stars at night. Eržvilkas is 200 km (125 miles) from the capital Vilnius, westwards along the E85 highway via Kaunas.

The Netherlands

Although the Netherlands has the reputation of being a very light-polluted country, there are islands (literally) of darkness to the north. In the West Frisian chain of islands, on the eastern side of the island of Terschelling, is De Boschplaat, an area of 40 km^2 (15 square miles). De Boschplaat became an IDA International Dark Sky

Park in 2015. No artificial lighting is allowed here, a gesture no doubt appreciated by the many European migratory bird species pausing on the island on their way to Africa. The area is one of the European Council's Natura 2000 Nature Reserves. Staatsbosbeheer, the local conservation agency, manages the site and, except when conservation work is in progress, allows unhindered public access. Eight sites within the park have been designated for astronomical observations (Fig. 3.65). "Because of De Boschplaat," says Wim Schmidt, leader of the effort to secure IDA recognition for the new Dark Sky Park, "Terschelling is one of the most visited tourist places in the Netherlands."

> darksky.org/de-boschplaat-named-first-ida-international-dark-sky-park-
> in-netherlands

Norway

More than 40 % of Norway lies above the Arctic Circle. In summer, the midnight Sun (in northern Norway) and very short nights in the south will curtail star-watching. In winter, a view of the aurora is an increasing possibility the further north you go. At the latitude of Bergen (60.4°N), southern Norway's enormous, glaciated Hardangervidda Plateau, with its National Park, is one of the country's

Fig. 3.65 Auriga (lower left), Perseus and Cassiopeia from Terschelling. (Photo: Wim Schmidt)

many starry areas. In the 1820s, Norwegian geophysicist and astronomer Professor Christopher Hansteen (1784–1873) chose the Hardangervidda for his stellar observations, and the sky there is still just as dark. The plateau is a rolling, treeless moorland with countless lakes and streams, flattening out towards the east with more vegetation. Go prepared, and take expert advice—challenging terrain and weather await you.

  en.hardangervidda.com
  www.nasjonalparkrute.no/eng

Portugal

Portugal has its skyglow, as the tourist trade populates the coastline and Lisbon continues to grow. Darker spots can be found: the widely known COAA Astronomy Center (Centro de Observação Astronomica no Algarve) near Portimao in the Algarve region of southern Portugal gives visiting amateur astronomers a chance to work with four sizable telescopes, the largest a half-meter diameter reflector, under good night skies (Fig. 3.66). Its southerly latitude (37°N) in Europe means that objects in constellations such as Scorpius and Sagittarius, low to the southern horizon from the UK or the northern United States, are much more accessible here.

Fig. 3.66 COAA domes. (Photo courtesy COAA)

These constellations contain some of the richest starfields, clusters and nebulae within the Milky Way stream. At this latitude summer nights are also longer and darker than they are further north.

www.coaa.co.uk

Poland

In March 2013, at the municipal center in the town of Lutowiska in the extreme southeastern corner of Poland, the Bieszczady Starry Sky Park, in the Bieszczady Mountains, officially came into being. This is now Europe's second largest dark sky protected area, and it is a pristine place beneath primeval night skies. Its literature boasts that "a clear night here will give you more than 7,000 stars, which is enough for anyone; you feel like you can reach out and touch the sky." A light limitation scheme was put into place early on in the dark sky process.

www.polska.pl/en/tourism/nature/under-stars

Romania

Many eastern European countries retain intrinsically dark areas across much of their terrain, and there are plenty of places in rural Romania where the starry sky is relatively unspoiled.

The Apuseni Natural Park, in Transylvania in western Romania, has some of the darkest sites for skywatching in the country. This mountainous area is known for its forests, meadows, alpine plant varieties and extensive cave systems, and has a thriving tourist infrastructure. The Natural Park, on the central northern flanks of the Apuseni Mountains, is bisected by Route 763, and there are well established hiking trails. Some of the area's guest houses (*pensiuni*) and hotels advertise their night sky views and welcome astronomers.

www.ruraltourism.ro/apuseni/pensiuniapuseni

Slovenia

On August 30, 2007, after much discussion with Dark Sky Slovenia, the Slovenian authorities adopted a light pollution law considered, at the time of writing, the strictest in Europe. DSS' Andrej Mohar summarized it in September 2007 thus: "The Law requires luminaires shining at 0 % above the horizon, which is stricter than the law of the Lombardy Region in Italy. The Slovene law allows 0.00 cd/

klm – in other words, absolutely no light above the horizon." For further details of this welcome legislation see www.darkskiesawareness.org/slovene-law.php.

In the mountainous northwest of Slovenia is Mount Triglav, 2864 m (9396 feet) high and the tallest peak in the Julian Alps and in Slovenia itself. Triglav appears on the Slovenian coat of arms and has been called "the pre-eminent symbol of the Slovene Nation." It gives its name to the Triglav National Park, Slovenia's only National Park. Organized stargazing takes place here, involving a cable-car journey. This is based around the new Bohinj Triglav National Park Center in Stara Fužina, with the participation of the Vogel Ski Center.

> www.bohinj.si/en/news/stargazing_bohinj

Spain

There are a few scattered pieces of Europe far away from the main landmass, and one of these that hosts several large international observatories is the 2423-m/7900-foot summit of La Palma, in the Canary Islands, where you can watch the Sun disappear over the Atlantic horizon as a bed of clouds a thousand feet below drifts in from the west. The Spanish territory of the Canary Islands is one of the world's astronomical hotspots, and on La Palma, at 28°N, giant telescopes of several nations, far away from the light spill of cities, drink in photons beneath matchless skies.

Negotiate the scores of hairpin bends on the road up to the top of the Roque de Los Muchachos (Rock of the Lads) and see a sky so full of myriad points of light that you might have the same problem as my school party had—the constellation figures are drowned out by the sheer numbers of visible stars crowding the heavens. Vast detail can be seen within the Milky Way, the Andromeda Galaxy stands out, and the zodiacal light is a solid presence. We stayed in one of the many rentable villas on the island. The main town is Santa Cruz de la Palma, and the scenery everywhere is stunning, especially in the giant volcanic caldera that occupies the northern section of La Palma. Of particular interest is the way the vegetation changes from tropical through temperate to tundra as you gain altitude on the way to the summits and their stars.

> www.ing.iac.es/PR/visits (observatory visits)
> www.islalapalma.com/en (general tourism/accommodation)

Sweden

The best night skies in Sweden are of course far from the towns of the southern and coastal areas of the country. Those staying in the south might be interested in the activities organized by the observatory and planetarium at Lund, a town with a heritage of five centuries of astronomical observing and research. The observatory offers public stargazing sessions and the planetarium (www.vattenhallen.lth.se/english/the-planetarium) gives regular shows.

About 200 km (125 miles) north of the Arctic Circle, in Lapland, not far from the border with Norway and Finland, is Kiruna, which is gaining a reputation as a tourist destination for the hardier breed of tourists. Ten miles from Kiruna in the village of Jukkasjärvi is the famous ice hotel, which is a good base for auroral observations. The Esrange Space Center (www.ssc.se/esrange), 40 km (25 miles) northeast of Kiruna, is involved in satellite design and rocket and stratospheric balloon launches. There are plans for Esrange to become Sweden's spaceport one day, but don't expect trips upwards from there soon. The facility itself is not open to the public, but its website mentions "a small unmanned visitor center with an interactive exhibition." Check that this is open before making the trip.

> www.icehotel.com/Winter
> www.kirunalapland.se/en/news/ssc-esrange-visitor-centre

Switzerland

There are of course many places in Switzerland where the night sky can be seen at its best, with mountains shielding the observer from any waste light emitted by towns in the nation's valleys. Perhaps the best of many examples of Swiss Alpine hotels where the stars may be seen on clear nights is the Kulmhotel Gornergrat, 3100 m (10,170 feet) above sea level, near Zermatt and close to the famous 'fang' of the Matterhorn. The most striking features of the building on photographs are its observatories, set atop two stone towers. Skiing is far and away the most popular pastime here, but the observatories, "equipped with everything a studious astronomer might need, but which, with a little guidance, even newbies can use," ensure that the Kulmhotel is also well patronized by amateur astronomers.

A list of Swiss astronomical organizations and societies appears on astronomy-events.ch/en/societies.

> www.myswitzerland.com/en-gb/accommodation/hotels.html

Turkey

Turkey still retains plenty of dark places, many of them along an axis between Sinop on the northern (Black Sea) coast and Anamur in the south, on the Mediterranean across from Cyprus. The coastal road to the west of Anamur looks a good prospect on the light pollution maps, as does the region of Tosya in the north.

In the central Anatolian province of Nevşehir, the historic and geologically unique region of Cappadocia, famed for its 'fairy chimney' rock pillars, has various places suitable for starwatching in the vicinity of the Ihlara Valley, Gulsehir and Kaymakli.

"In the center of the Pacific" might be the best answer to this question, but sites on land often mentioned in this debate are the Atacama Desert in Chile and the mountaintops of the Canaries, unsurprisingly chosen as sites for some of the world's greatest telescopes. Also mentioned are the 'Red Center' and the Nullarbor Plain in Australia (Fig. 3.67), and great swathes of Asia from Tibet through Siberia.

Ten years ago, astro-tourism was a rarity in the Atacama Desert, even though its many clear nights and absence of light pollution make it a Bortle Class 1 stargazers' 'Nirvana' almost everywhere. Some of the world's greatest astronomical facilities have taken up residence here in this high-altitude rocky expanse, which astronomers have likened to the surface of Mars. Some of the observatories have been included in organized astro tours. A brief Internet search will reveal what's on offer in this region. This is a high plateau desert, and most days are clear, though temperatures can be low, and at night below zero in winter months. From January into March, the so-called Altiplanic winter brings occasional showers. Contrary to popular belief, the Atacama is not totally without rain. Only in certain areas is it unknown within living memory.

The glittering southern constellations and the wonderfully bright Milky Way over Australia, unfamiliar to many readers of this book, are beautifully captured on

Fig. 3.67 Southern stars from the heart of Australia. The two Magellanic Clouds, satellite galaxies of the Milky Way, are prominent. (Photo: James Hilder)

many websites. For a flavor of the night sky in dark southern places, check out www.wired.com/2011/07/time-lapse-night-sky, featuring a striking video by Alex Cherney; or www.youtube.com/watch?v=UmEC7goXBII by Phil Hart and others.

A real bonus for travelers to the south are the two Magellanic Clouds, the Milky Way's smaller companion galaxies (well seen in Fig. 3.67). Seasonal weather variations in Australia are less distinct than in the Northern Hemisphere, and average annual rainfall in Australia's arid desert is low. Baking heat in summer is connected to dust storms, thunderstorms and occasional bush fires. For recommendations on the best times to visit Australia's dark desert places, see www.outback-australia-travel-secrets.com/outback-climate.html.

There are no International Dark-Sky Association Reserves, Parks or Communities in central Asia, but a glance at the light-pollution maps reveals that much of the region doesn't really need any. From the Indian border up towards the Siberian Arctic, darkness rules. In Mongolia, for example, away from Ulan Bator (now increasingly spelled Ulaanbaatar) where over a million people live, lighting is sparse. An Internet search for "Mongolia dark skies" will give information about the quality of the stars over the country and both organized and informal trips to see them. Mongolia's climate can be summarized as extreme Continental—warm, sunny summers, hence its reputation as "the land of blue skies," and long, dry and sometimes bitterly cold and snowy winters. Indeed, the average temperature in most of Mongolia is below zero from November to March, and not much warmer from April through October.

Wherever you go, have a safe and cloud-free stay at your chosen dark sky site!

Chapter 4

What's in the Sky?

In silence we watch as, from behind and above the silhouettes of ponderosa pines swaying against the darkest blue-black sky, emerge dozens, then hundreds, then thousands of stars.

–Paul Bogard, The End of Night (2013)

If you're fairly new to astronomy, looking up at a dark night sky you're unfamiliar with can be a daunting experience. You'll see huge masses of twinkling dots (Fig. 4.1)—and some that don't twinkle (probably planets). Can't make out a lion, a dolphin or a bear, or a mighty hunter with a belt—ah, but look over there, three bright stars in a line. Might that be Orion's Belt?

You've taken the first step. You've found something. Could be easier if you had some help. If there's no informed amateur astronomer at hand, there are phone apps, planetarium programs that guide you around the starry vault, countless websites, books, YouTube tutorials…

There is, however, a lot you can do and a lot of enjoyment to be had just looking, even if you don't know the stars well. Be patient. After a while you're likely to see the half-second flash of a meteor burning its way through the upper atmosphere—or is it an Iridium communications satellite reflecting the light of the Sun?

From a dark place, you can locate the Milky Way, if it's at a suitable angle in the sky, without much trouble—a broad, mottled river of faint light, divided by a dark rift into two distinct streams towards the southern horizon. If the Moon's around, even a modest binocular will reveal a few of its craters and some detail of the dark patches, basaltic plains that olden-day astronomers thought were seas—and we still call them that (Fig. 4.2). Did I write 'binocular'? Officially, a pair of binoculars is two instruments, but almost nobody calls binoculars a binocular, so binoculars they will be from now on in this book.

© Springer International Publishing Switzerland 2016   155
B. Mizon, *Finding a Million-Star Hotel*, The Patrick Moore Practical Astronomy Series,
DOI 10.1007/978-3-319-33855-2_4

Fig. 4.1 Crowds of stars above a Scottish aurora. (Photo: James Hilder)

Suddenly, you have the great good fortune to see faint red patches forming in the sky, and a greenish glow rising low in the north. Even in mid-latitudes, people out at night see the northern lights (aurorae), and more often than some would expect.

With a good chart, a little experience and a deal of patience—and, of course, clear weather—you may soon be finding some of the objects mentioned below, and experiencing that thrill of discovery as light that's been traveling for tens or thousands or millions of years enters your eyes after its unimaginably long journey through the void.

No telescope? Contact local astronomical groups or ask astro-tourist destinations if they have instruments on loan. Are there local experts who hold regular telescope sessions for visitors? If there are night-sky objects you'd like to see, they will be only too pleased to find them for you. Some destinations will have equipment for astrophotography, and arrange for visitors to take their own photos through telescopes. You can even secure passable 'snaps' of the Moon simply by holding a phone equipped with a camera to the eyepiece of a telescope. With the right positioning of the camera lens against the eyepiece, and a steady hand, you should get results. This author's photo (Fig. 4.3) of the Moon was taken with an I-phone through a 12-inch reflecting telescope at a star watch organized by the Wessex Astronomical Society (Wimborne, England) in 2014.

First in this chapter, some daytime astronomy—our Sun and the phenomena, both atmospheric and more distant, associated with it. Then, into the night—the sky season by season with basic charts and notes, zeroing in on some relatively nearby

The Moon aged 9 days, 2013 February 19 23:36–23:38
Celestron 100mm ED refractor f9, DMK41AF02.AS camera, Baader 685nm IR filter.
Processed in Austostakkert 2, Registax 6, DoubleTake 2 and Photoshop CS4.
David Arditti, Stag Lane Observatory, Edgware, Middlesex, UK

Fig. 4.2 The 'seas' of the Moon. (Photo: David Arditti)

stars, a selection of the Sun's stellar neighbors. Next, the Moon, then the planets of the Solar System, before moving further into our own Milky Way Galaxy of half a trillion stars. We encounter some of the galaxy's better known objects, such as

Fig. 4.3 The Moon captured with an iPhone though a 12-inch telescope. (Photo: Bob Mizon)

clusters and nebulae. Finally, ever outward to some of the more remote galaxies you might see through powerful telescopes.

Many of the images in this book were kindly donated to this project by talented astrophotographers in both the UK and the United States, and they show things as you'd *never* see them through a telescope or binoculars. The gaseous wreaths of the Orion Nebula and the intriguing annulus of the Ring Nebula may glow with wonderful colors in the illustrations; but when you view them through the optics, expect extended night-sky objects like these to be gray! Individual stars can also show wonderful colors, and we'll discover some of these later. But let's start with the most important star of all. It's the one that drives our weather, feeds us and moves our vehicles.

Our Daytime: The Sun and Safe Observing

The Solar Disk

Any worthwhile discussion of observing the Sun starts with a very important warning: **Never, never look directly at the Sun**, with or without optical aid. Anyone peering through an ordinary telescope or binoculars at the Sun will be blinded immediately,

Fig. 4.4 Sun safety: eclipse glasses. (Photo: Bob Mizon)

and for life. Looking at it for only a second or two with just the unprotected eye can cause significant retinal harm. Your eye is an extension of your brain, and brain cells aren't replaced when damaged.

During total eclipses, before the Sun disappears behind the Moon, enthusiastic onlookers have injured their eyes by using 'traditional' impromptu filters such as candle-smoked glass, tinted glass, sunglasses or even colored cellophane. About the only effective screen available in everyday use (in industrial units nowhere near most of us) is a Number 14 welder's mask. Safe solar film (for example, by Baader), commercially produced filters or certified eclipse viewers/glasses (Fig. 4.4) are best, and all are easily found online or can be obtained from astronomical organizations or telescope dealers. Make sure they are absolutely clean, with no tiny holes or scratches. Never look through telescopes or binoculars while wearing eclipse glasses. They are for direct viewing only.

There are still telescopes around that come with a small solar filter, often marked SUN. This, says the manufacturer, is screwed onto the eyepiece for eye protection when viewing the solar disk. If you have one of these, the best course of action is to destroy it. These potentially deadly little filters heat up as sunlight is focused on them, and have been known to crack and separate—and you don't have time to get your eye away before it is permanently injured.

Fig. 4.5 Projecting the Sun's image during the total eclipse of August 11, 1999. (Photo: Steve Tonkin)

With a telescope or binoculars on a tripod, you can project the Sun's light through a hole onto a screen or into a shoebox—an image of the Sun produced like this can be remarkably detailed (Fig. 4.5).

Above, we mentioned not looking at the Sun with an *ordinary* telescope. So is there an *extraordinary* one that's safe? Are there safe solar filters?

Take expert advice before embarking on any of the following. You can safely see the Sun in some detail through telescopes by firmly fitting a solar filter over the *objective end* of the telescope (the end towards the Sun). This blocks ultraviolet and infrared radiation (heat) from the Sun and nearly all its brightness, so the inside of the instrument doesn't heat up much. Such filters are made from aluminum-coated Mylar, or specially coated glass—but do get advice from a reputable dealer.

If you are using a telescope with a finderscope (Fig. 4.6), remember that the finderscope also focuses the Sun's light,so make sure its caps are on, or even better, remove it altogether. If using binoculars to project the Sun, cap one of the objective lenses and use one half of the instrument only.

Specialized solar telescopes are becoming ever more popular with daytime astronomers. These are often called PSTs, Personal Solar Telescopes (Fig. 4.7). They admit only one wavelength of light, the most active emitted by hydrogen atoms. They reveal features such as prominences, which are outbursts many times the size of Earth, and filaments, sinuous darker lines on the Sun's surface, actually prominences seen from above (Fig. 4.8).

Fig. 4.6 A finderscope sits atop Les Fry's reflector. (Photo: Bob Mizon)

Fig. 4.7 A specialized solar telescope. (Photo: Ron Westmaas)

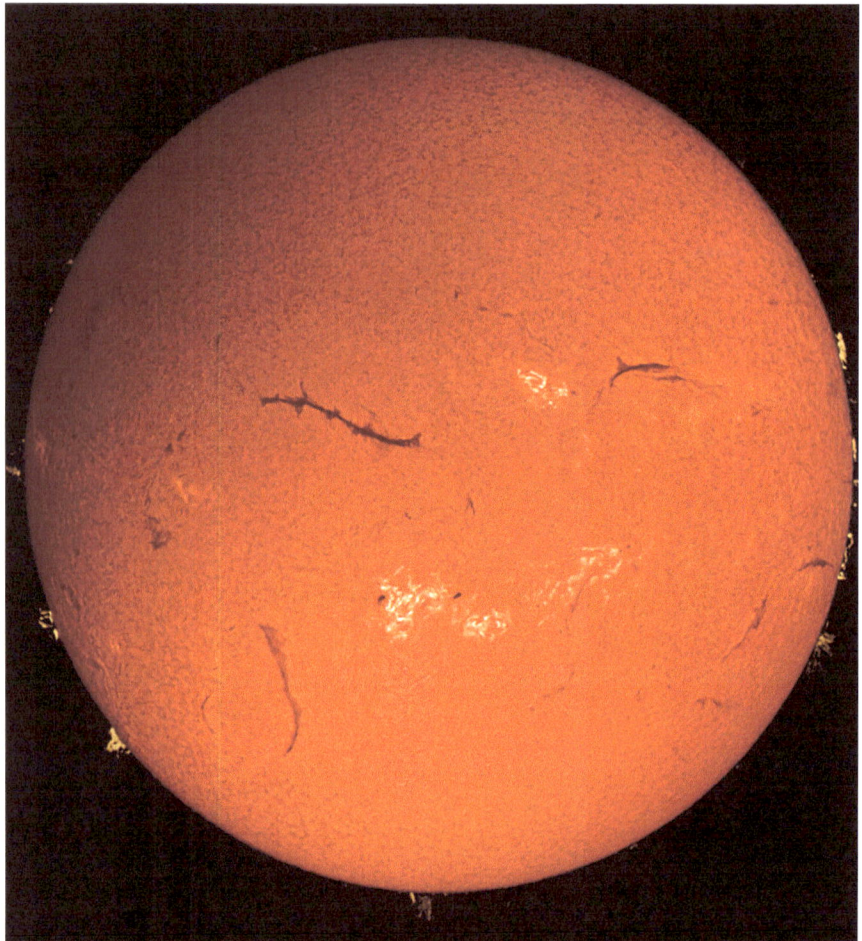

Fig. 4.8 The active surface of our Sun. (Photo: Sheri Lynn Karl)   .

The Aurora

Energized material ejected from the enormous furnace of the Sun is funneled by Earth's magnetic field towards its North and South poles. In the north, the resulting display is the aurora borealis (Fig. 4.9), and in the south, the aurora australis (southern lights). The active area spreads away from polar regions to form an auroral oval around the globe that, if the Sun is particularly active, may descend to mid-latitudes, and in exceptional cases to the equator. Find the current location of the oval on the informative www.spaceweather.com, or for Alaska, on www.gi.alaska.edu/AuroraForecast.

Fig. 4.9 A magnificent aurora. (Photo: Paul Curtis)

This author has lived in his present house on the southern coast of England, at latitude 50.5°N, since October 1982, and has seen the aurora from the back garden about 20 times. How can this be? My incredulous neighbors have never seen it, and some have lived here even longer. The reason is obvious—they don't look for it. On most clear nights I step for a short while out of the back door, allow a little time for my eyes to adapt to the dark, and scan the northern sky. It helps that, even though I live near a big town, I'm on its northern edge and, praise be to the local authority, the streetlights around me point strictly downwards and go off from midnight until 5 o'clock in the morning. I remember vividly the night of March 13, 1989, when massive auroral activity spread across the northern and southern hemispheres. As the evening progressed, the sky, lit at first with a green curtain across the north, became ablaze with vivid reds, greens and subtle bands of blues and yellows, producing silvery-white vertical rays. The display gradually spread southwards, and filled the sky completely. Enormous wedges of rosy light climbed from the northern horizon (Fig. 4.10), and luminous streamers snaked above at unexpected speeds. You can get lucky!

The Great Aurora of 1989 was exceptional, a once-in-a-lifetime event. But with patience, a clear northern horizon and little or no light pollution, you might one night see a green arc climb slowly from the northern horizon (Fig. 4.11), or faint rays rising—the first tentative fingers of the aurora borealis. Aurora is the Latin word for "dawn," as the phenomenon was thought to resemble the dawn breaking in the north instead of the east. Boreas was the Latin name of a Greek north-wind deity (Βορέας).

Fig. 4.10 The major auroral display of March 13, 1989, from Wimborne in Dorset, England. (Photo: Bob Mizon)

Fig. 4.11 The rising green arc that precedes many auroral events. (Photo: Iain Cartwright)

Rome is at 42°N, and the ancient Romans saw the aurora's red glow as a rare and bad omen. In AD 37 the Rome fire brigade hurried to the port of Ostia to tackle a blaze that turned out to be the northern lights. History repeated itself in England in 1938, when a particularly bright display brought firefighters to the royal residence of Windsor Castle.

How high up in the sky will your aurora be? It's easy to estimate altitudes and distances in degrees in the sky. From horizon to horizon, it's 180° (though you're likely to have trees or other objects on your horizons). So to the zenith, vertically above you, it's 90° from the horizon. For smaller angles, no equipment is needed apart from your hand: with arm outstretched, and hand extended and splayed, the distance between the tips of the little finger and thumb is about 20°. A clenched fist is about 10° wide. The first three fingers held together are about 5° across. Your thumb covers two degrees, and your little finger one. The diameter of the full Moon is half a degree—surprisingly small. You can more than cover it with your little finger.

The aurora has been interpreted throughout history in many different ways. Some Northern European aurora legends can be found on a website created by a Swedish physicist, the late Professor Ingrid Sandahl: www.irf.se/norrsken/Norrsken_history. html. Interpretations of the aurora from early North American legends can be found at www.indigenouspeople.net/aurora.htm.

Aurorae are seen on other planets: find striking images of auroral storms on Jupiter at apod.nasa.gov/apod/ap001219.html and on Saturn at apod.nasa.gov/apod/ap990123.html.

You could go to the trouble expended by the Geophysical Institute of the University of Fairbanks, Alaska, and install a 360-degree all-sky video camera to catch these intriguing displays (allsky.gi.alaska.edu), but the simple method described above (look north, dark-adapt, be patient) should suffice. The aurora is not difficult to photograph. A tripod-mounted camera operating at various exposures (from a couple of seconds up to 30 s) will capture its beauty if the sky is reasonably dark. An exposure of a light-polluted sky tends to wash out anything fainter. Professional photographers such as Oliver Taylor (Dorset, UK) have taken striking images of the northern lights (Fig. 4.12).

Sundogs, Arcs and Crepuscular Rays

A common sight in the sky when there is mixed sunshine and thin cloud are sundogs (parhelia), resulting from refraction of sunlight within plate-like ice crystals floating high up.

Many people fail to notice them. Whenever there are Sun rays and the right kind of thin cloud around, you may observe (not while driving!) what look like miniature rainbows hovering about 22° to the left and right of the Sun (Fig. 4.13), with the red on the sunward side. Bright ones are easy to photograph. Less often, when cloud conditions are right, there may also be small colored parhelic arcs above the Sun, or even a striking full halo with both sundogs (Fig. 4.14). One very basic solar

Fig. 4.12 Auroral rays over a poppy field in Dorset. (Photo: Oliver Taylor)

effect is that of crepuscular rays (Fig. 4.15). These are the shadows of clouds thrown onto other, usually thinner clouds, creating beautiful patterns of radiating spokes centered on the Sun.

Fig. 4.13 A parhelion (sundog) from the author's back door. (Photo: Bob Mizon)

Sun Pillars

Sun pillars, also known as light pillars, are columns of light towering above the Sun when it is low to the horizon. They are more common in cold weather and are caused by light reflected upwards and downwards from flat ice crystals. The crystals that cause sun pillars consist of flat hexagonal plates, all oriented more or less horizontally as they fall through the cold air. They can be a remarkable sight, looking like a giant exclamation mark with the Sun as the 'dot' at its foot (Fig. 4.16).

Fig. 4.14 Full solar halo, Wards Island, Toronto. (Photo: Mark Trusz)

Fig. 4.15 Crepuscular rays radiate across the sky above Puddletown, Dorset. (Photo: Chris Bowden)

Fig. 4.16 Sun pillar, Lairg, Scotland. (Photo: Graeme Whipps)

Noctilucent Clouds

Noctilucent clouds (NLC) move in a slow, rippling dance through the high atmosphere (Fig. 4.17). They appear as interlaced or ribbed skeins of pale silvery blue light and are seen in twilight skies, when the Sun is not far below the horizon in summer. Their composition is not well known, but they are thought to be particles interacting with ice. The lack of reports of NLC before 1885, when they were thought by some to be connected with the enormous eruption of Krakatoa 2 years earlier, and the increasingly frequent sightings since, suggest that they may be caused by a mixture of volcanic ash and products of industrial pollution in the upper reaches of the atmosphere. They may well be indicators of climate change. A superb video of advancing NLC by UK-based astrophotographer Richard Fleet is at www.popastro.com/news/newsdetail.php?id_nw=185.

The Zodiacal Light

Before wasted light from distant cities and local lamps colored our skies, the zodiacal light (Chap. 1, Fig. 1.11) must have been a common sight in a starry, moonless sky. It takes the form of a faintly glowing cone, slanting upwards from

Gloworm and NLC   Dave Tyler

Fig. 4.17 A glow-worm in the grass fails to admire noctilucent clouds in the northern sky. (Photo: Dave Tyler)

its base along the horizon. It appears in the east shortly before dawn or in the west shortly after dusk, and is at its best on March evenings and in September dawns when the ecliptic (the path of the Sun and planets in the sky) is at its steepest. In a really dark place, it can rival the Milky Way in brightness. The zodiacal light has been seen by the author boldly tinting an unpolluted velvet-black sky from the top of the nearly 8000-foot Roque de Los Muchachos on La Palma, Canary Islands, and from Brockington in rural Dorset, in the Cranborne Chase AONB mentioned in Chap. 3. The subtle glow of the zodiacal light is the product of sunlight scattered from countless dust particles left behind by vast numbers of comets looping through the Solar System.

Nighttime Astronomy: Wonders Without End

Constellations: Stars, Science and Myth

In an unpolluted sky, sharp-eyed observers may see up to 7000 stars. The pale celestial river of the Milky Way stretches across the celestial vault for much of the year, and late at night in summer in the Northern Hemisphere it arches high above.

Its countless faint stars, so distant and condensed that they look like a continuous flow of milk or ash in the sky, remained undetected until Galileo turned his small telescope on them in 1610: "I have observed the nature and the material of the Milky Way… innumerable stars grouped together in clusters… arranged in a wonderful manner." Along the sky-river, and further afield away from the plane of the Milky Way, brighter stars shine out to form constellations. Some, either relatively near to us or intrinsically super-bright, are glittering beacons, standing out from the crowds. Blue-white Vega, in the constellation of Lyra, the Lyre or Harp, glitters high above in August. The silvery spark of Deneb, in Cygnus the Swan, draws the eye towards a mighty rift of dust, the ashes of long-dead stars, dividing the northern Milky Way.

Some scientists think that there are more stars in the observable universe than grains of sand on every one of the world's beaches. If this is the case, then why don't we see stars in all possible directions, causing the sky to be uniformly bright? Heinrich Olbers (1758–1840), a German physician and amateur astronomer, applied his mind to this conundrum, ever since known to science as Olbers' paradox. The paradox is resolved chiefly by the fact that the universe is expanding (and, according to recent research, also paradoxically accelerating as it does so!).

The ancients' view of the universe as static and unchanging is no more. We now realize that the attenuation and reddening of the energy of distant starlight, caused by the enormous velocities of the recession of the fleeing galaxies, rob us of much of the light of remote astronomical objects. The darkness of the night sky is mainly due to what E. R. Harrison of the University of Massachusetts eloquently called "the infrared gloom of the Big Bang." The clouds of dark dust within our own galaxy also soak up starlight. These vast billows of carbon and silicon are well seen in photos of the Milky Way, for example Fig. 4.18 by accomplished Welsh astrophotographer Nick Hart. It's not just light pollution that has stolen our stars—dust and the flight of the distant galaxies also play their part.

We may have been told in school that the ancient Egyptians, the Romans or the Babylonians created the constellation figures: dot-to-dot representations of heroes, gods, animals and legendary characters. Some of them date back much further than that, however. Some 17,000 years ago, long before those civilizations flourished, an artist balancing on a crude ladder or platform in the depths of a cave in southwestern France painted an image of haunting beauty high up on a wall. The Bull of Lascaux (news.bbc.co.uk/1/hi/sci/tech/975360.stm), a depiction of one of four mighty aurochs bulls, lost for millennia along with 2000 other cave paintings, was discovered in 1940 by Marcel Ravidat, a teenager searching for his lost dog. Around the lunging bull, the artist, perhaps the first known amateur astronomer, has added the most memorable features of the area of the night sky occupied by our celestial Bull, Taurus. The crowd of dots shown above its back are the cluster of the Pleiades or Seven Sisters; the dots in its head represent the Hyades cluster, which forms the Bull's head in the sky; the bright star we call Aldebaran is the eye. The Pleiades and Hyades are well seen in Chris Bowden's

Fig. 4.18 Dark clouds mingle with bright nebulae around the star Sadr in Cygnus. (Photo: Nick Hart)

photo (Fig. 4.34) of a shooting star below the V-shaped head of Taurus. In front of the bull on the cave wall is a line of dots, four in number, placed to represent the Belt of Orion with an extra star that may represent Sigma Orionis, just beneath the left-hand belt star in the sky. That imagined bull has been part of celestial lore for a very long time.

The 48 classical constellations were recorded in great detail, star by star, by Ptolemy of Alexandria in AD 150. Now, with the addition of southern constellations, there are 88 sky figures in the Western tradition. Indeed, there used to be more, but some, like Felis the Cat, Lumbricus the Earthworm (!) and Solarium the Sundial are long gone.

These brief notes and the following basic seasonal sky charts kindly prepared for this book by Wessex Astronomical Society astronomer Alan Jefferis refer to the night sky over northern mid-latitudes. For greater detail and views further south in the sky, as seen from Florida, for example, see www.skymaponline.net.

Chart 1 The Stars of Spring

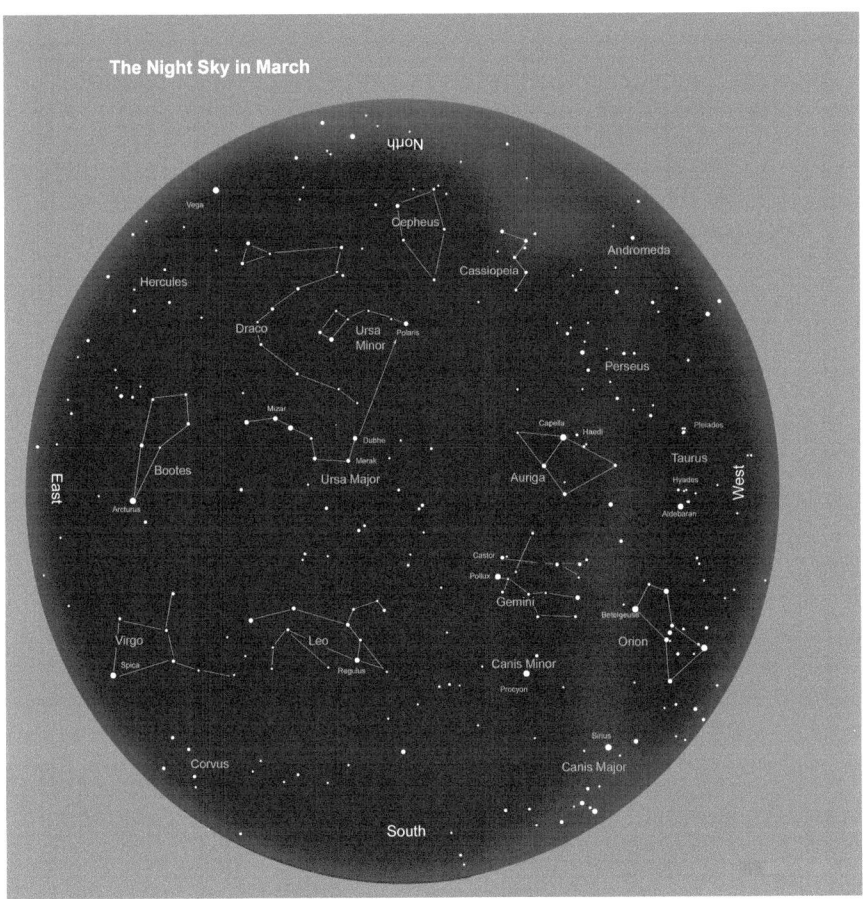

(March 30, 10 p.m.)
The Milky Way lies across the western night sky around 10 p.m.

IN THE SOUTHERN SKY. The last traces of the winter constellations now fall to the west: Orion, Castor and Pollux (in Gemini, the Twins) and red Aldebaran in Taurus the Bull linger as night falls. In the south, the distinctive hook of the Sickle of Leo, the Lion, stands upon Regulus (77 light years away). The very large constellation of Virgo, the maiden of spring, now appears above the eastern horizon, announced by Spica, representing the stalks of wheat in her hand. Spica is 260 light years away, so the light we see now began its journey at about the time of the famous

1758 return of Comet Halley, predicted by the great astronomer decades earlier. The seven stars of the Plough (Big Dipper), in the Great Bear, Ursa Major (Fig. 4.19), are now high above in the late evening.

Fig. 4.19 Bears in the sky over Loch More cottage. (Photo: Gordon Mackie)

In the Northern Sky. In the northwest glitters yellowish Capella, in Auriga, the Charioteer. It is the northernmost first-magnitude star. Three little stars near Capella, which means "the Goat," are called the Haedi ("Kids"). Vega, 25 light years away, is the very bright object rising in the northeast. Around April 16–25, the swift, whitish Lyrid meteors appear to emanate from down near the northeastern horizon. The unmoving Pole Star in Ursa Minor, the Little Bear, is the infallible marker of the north, and all the other stars circle around it anti-clockwise at a rate of one degree every four minutes. Find it by using the Plough/Big Dipper, whose two end stars Merak and Dubhe point towards it (Fig. 4.20). In the northwest is the distinctive "W" shape of Queen Cassiopeia, mother of Princess Andromeda, with Perseus, Andromeda's rescuer. These three constellations are well worth sweeping with binoculars for their glorious star clusters.

Fig. 4.20 The two right-hand stars of the Plough/Big Dipper, Merak and Dubhe (bottom center) point to the Pole Star (brightest star, top center). (Photo: James Hilder)

Chart 2 The Stars of Summer

(June 30, 10 p.m.)
The Milky Way lies across the eastern night sky at around 10 p.m.

IN THE SOUTHERN SKY. Nights become frustratingly short by midsummer, but at least freezing temperatures are unlikely. By midnight, the Milky Way soars majestically above, from the southern point of the horizon to the northern for observers under dark skies. A moonless night is essential to see the Milky Way at its best. Can you detect the orange hue of the brightest star of the sky's northern hemisphere, Arcturus in Boötes? Scorpius rears its head low in the south, and the reddish hue of its brightest star Antares should be apparent if mists don't intervene. Libra, Virgo and Leo lie on the zodiac, the path followed by the Sun, Moon and planets through the sky.

IN THE NORTHERN SKY. Embedded within the Milky Way in the northeast are Cassiopeia, Cepheus and Cygnus the Swan with its bright beacon Deneb (1500 light years away). Deneb is part of the huge Summer Triangle, with Vega and Altair at the other corners. The Triangle will be a feature of the sky long after summer is over. On August 11–13, expect to see Perseid meteors rising from the northeast. This is probably the most reliable and productive meteor event of the year. The Plough/Big Dipper (in Ursa Major, the Great Bear) and Cassiopeia sit at either side of the Pole Star. Between them wind the coils of Draco the Dragon. Remember that the stars of the Great Bear are not just the glittering seven of the Plough/Big Dipper. This large constellation spreads far beyond this figure, and harbors many faint, distant galaxies for telescope users.

Chart 3 The Stars of Autumn/Fall

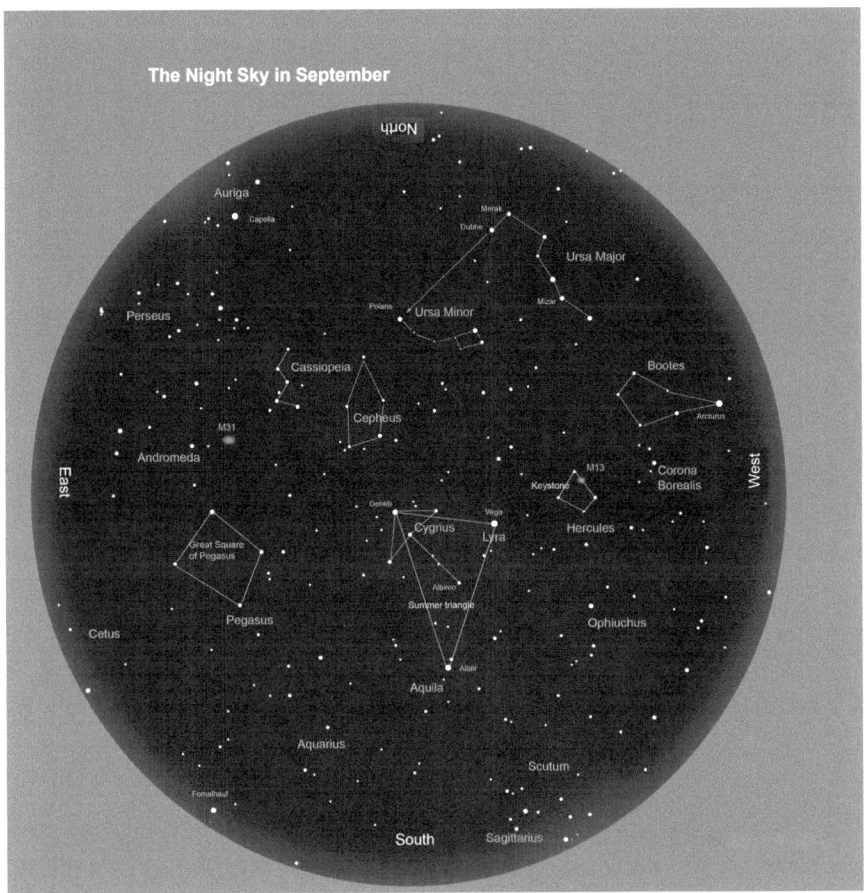

(September 30, 10 p.m.)

The Milky Way is high in the night sky around 10 p.m.

IN THE SOUTHERN SKY. The easily identified Square of Pegasus is due south around midnight in September. The constellation bears little resemblance to a flying horse. Not far from brilliant Vega, in the southwest, is the ancient figure of Hercules.

Fig. 4.21 M13, the globular cluster in Hercules. (Photo: Dave Finnigan)

None of his stars is particularly outstanding, but the "Keystone" of Hercules harbors a telescopic gem: the globular cluster M13 (Fig. 4.21), 25,000 light years away and compared by some to a pile of sugar on a black velvet cloth when seen through a large telescope. The Summer Triangle of Vega, Deneb and Altair sits high in the west. Sweep it with binoculars to reveal myriad clusters and lines of faint stars along the plane of the Milky Way. The oval blur of M31, the Andromeda Galaxy, is climbing in the southeast, and the stars of Sagittarius glitter in the south where the Milky Way meets the horizon. Sagittarius marks the direction of the center of our galaxy and its enormous black hole with a mass of 3 million Suns.

IN THE NORTHERN SKY. The Plough/Big Dipper descends toward the northern horizon. Queen Cassiopeia now rules in the northeast, enthroned on the Milky Way's starry stream and close to the line of bright stars of Princess Andromeda and the glittering clusters of Perseus. Capella, in Auriga, is rising in the late evening in the east, and by midnight, winter constellations such as Gemini the Twins and Orion (see Chart 4) will be peeping above the eastern horizon. On October 21–22 the Orionid meteors are seen. They are usually not as numerous as the summer Perseids or the winter Geminids.

Chart 4 The Stars of Winter

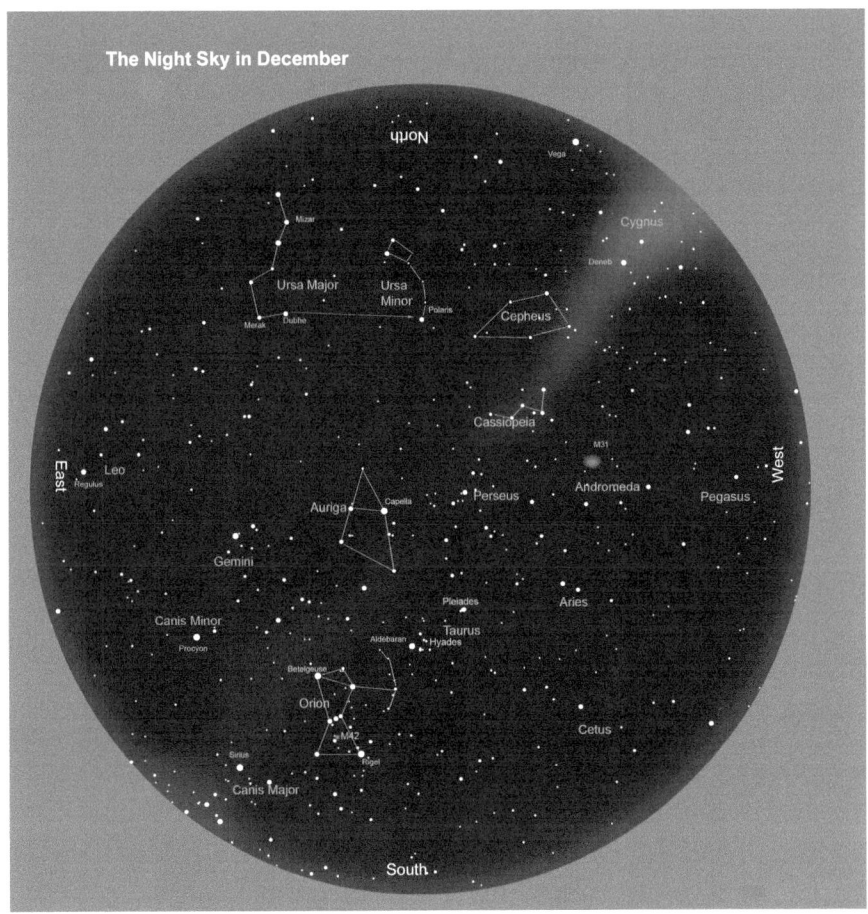

(December 30, 10 p.m.)
The Milky Way is high in the night sky around 10 pm.

IN THE SOUTHERN SKY. Orion dominates the sky in the south. Use optical aid to see detail in his nebula, M42, just below the central belt star. M42 is a mighty star factory 1300 light years away. Sirius (8.6 light years), in the sparkling constellation of Canis Major, the Great Dog, is the brightest star in the night sky. Its name is from a Greek word meaning 'blazing': it twinkles violently as its light passes through denser atmosphere. Below Auriga, Taurus the Bull, with its striking star clusters, attacks Orion from the west. The Geminid meteors flash on December 13–14, and often produce bright white trails. (See them at earthsky.org/space/everything-you-need-to-know-geminid-meteor-shower).

IN THE NORTHERN SKY. The seven stars of the Plough/Big Dipper are now down near the northern horizon. The three stars of its 'handle' were seen by some Native Americans as hunters pursuing the creature around the pole. The middle hunter is Mizar, a superb double star through a small telescope. Leo, with its bright star Regulus, now rises in the east, as Vega touches the western horizon. Look for the Leonid meteors soaring in the northeast on November 16–18. These often leave lingering gas trains behind them that can be studied with binoculars. Sweep the area of Cepheus the Ethiopian King with optical aid for sparkling star clusters and multiple stars.

Selected Stars: The Stories Behind the Twinkles

We will not try to describe the whole night sky in detail. Appendix C lists books, websites and planetarium programs that do that. Here are ten famous and easily found stars, with finder charts, that you can enjoy viewing up close from a dark sky location.

Chart 5 Albireo and Deneb finder chart

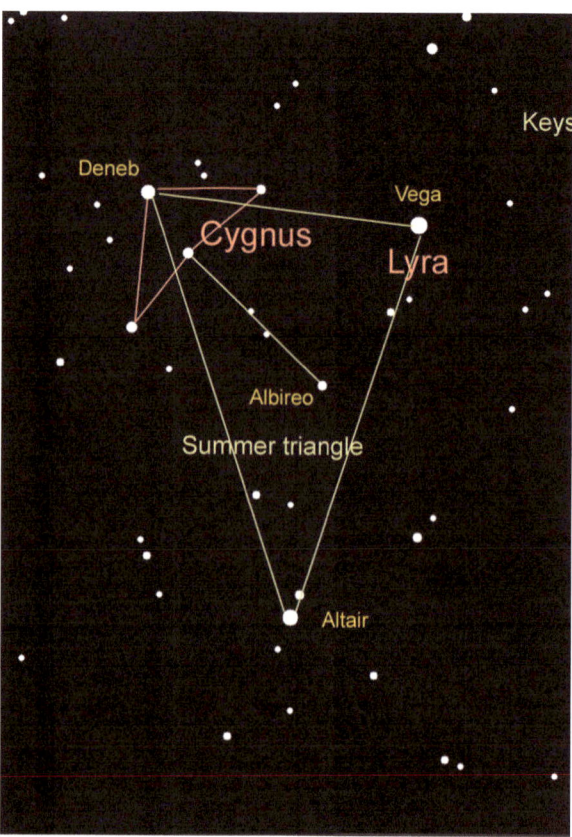

Fig. 4.22 The bright trio of the Summer Triangle over the ruins of St. Benet's Abbey, Norfolk, eastern England. Vega is the bright star above the doorway; Deneb, in Cygnus, is above Vega, and Altair is to its left. (Photo: Shaun Reynolds)

Albireo, the Summer Surprise

At the 'beak' of the celestial Swan Cygnus (Fig. 4.22), which appears to be flying along the river of the Milky Way in summer, there is an unremarkable-looking medium-bright star. It's Albireo, and, although with that *Al-* at the beginning it sounds like one of the many stars with an Arabic name (Aldebaran, Algol, Altair, for example, the *Al* simply meaning 'the'), nobody knows where the word comes from.[1]

When fledgling astronomers first see Albireo, through even a modest telescope, they are likely to exclaim with delight. It's not just a white dot, but reveals itself as a twin jewel. The brighter member of this celestial pair is of a golden honey color, while its slightly fainter companion glitters electric blue. Albireo is 430 light years

---

[1] For information about the fascinating and tangled history of star names try to find an absorbing if very old book: Richard H. Allen's *Star Names and Their Meaning* (1899), reprinted by Dover in 1963 as *Star Names – Their Lore and Meaning*. More recent is the *Dictionary of Modern Star Names* by Paul Kunitzsch and Tim Smart (Sky Publishing Corporation, 2006).

away. Its beautiful light left the star's surface at about the time when the first colony of Roanoke was established, and when Shakespeare, an aspiring young actor in his twenties, was seeking work in London. Albireo is the medium-bright star to the left of Vega in Fig. 4.22.

Chart 6 Antares finder chart

Antares, the Rival of Mars

Low in the southern night sky in June and July is the curving constellation of Scorpius, the heavenly Scorpion. For many mid-latitude observers, only the upper part of the Scorpion is seen, and the sting-star in its tail is below the horizon. Antares, the star at the Scorpion's heart, is bright and easy to find. It is a bloated, late-stage red giant, 800 times wider than the Sun and 553 light years away. Its average density is less than one-millionth that of the Sun. Sometime in the future it may well go supernova. Its red color long ago earned it the title 'Rival of Mars' (Mars in Greek being Ares). Train your binoculars or telescope on this star to enjoy its ruddy hue, then pan across a little to the right and you'll find a fuzzy ball of light looking like an out-of-focus star (Fig. 4.23). This is M4[2], a globular cluster 7000 light years away, one of more than a hundred that surround our Milky Way Galaxy.

---

[2] apod.nasa.gov/apod/ap000523.html.

Fig. 4.23 The glittering stars of the head of Scorpius (top right) over a Cornish beach. The bright, reddish star is Antares (right of center), and the smudge of light to its right is M4. (Photo: James Hilder)

Chart 7 Arcturus finder chart

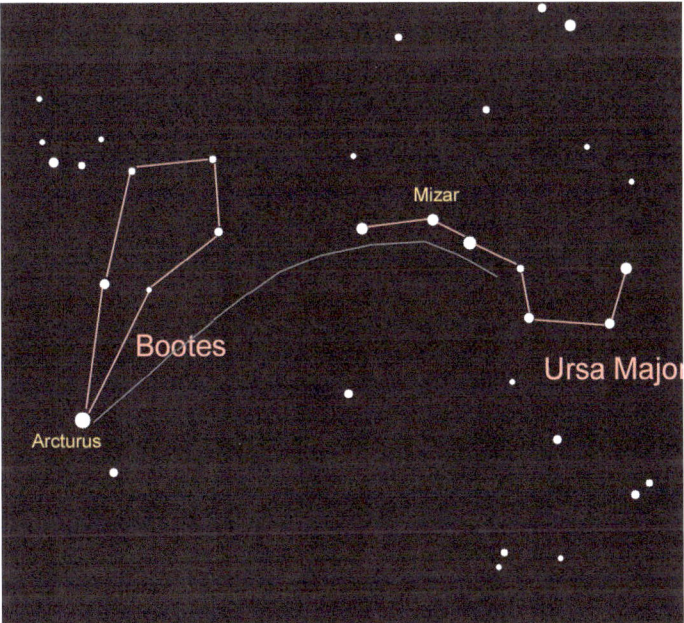

Arcturus, the Bear Driver

Mid-evening in the summer, look towards the west. A bold orange star dominates the sky, well above the horizon. This is brilliant Arcturus (below right of center on the summer chart), brightest jewel in the northern hemisphere of the sky. This aging giant's color reflects its relatively low surface temperature, a mere 4300°C! Arcturus, in the large constellation of Boötes, the Herdsman, appears to push the two bears, Ursa Minor and Ursa Major, around the polar point, hence its ancient name of 'Bear Driver.'

Arcturus' light has been traveling towards us for 36 years. Mythology apart, Arcturus has played its part in human affairs. In 1933, the illuminations at Chicago's Century of Progress Exhibition were switched on by focusing its light though telescopes onto photoelectric cells. The star had been chosen because the previous such event had been in 1893, 40 years before, and it was believed at the time that the star was 40 light years away. In the nineteenth century, astronomers at the Royal Observatory, Greenwich, first measured the heat from this star. Modern estimates suggest it is about the same as that from a candle 5 miles away. Through a telescope, Arcturus' orange hue is very obvious.

Chart 8 Betelgeuse and Sirius finder chart

Betelgeuse, the Next Supernova?

Another winter gem is one of Orion the Hunter's shoulder stars, Betelgeuse. This is often translated as the 'Hand,' or more surprisingly, the 'Armpit' of the Central One, or Giant. Betelgeuse is noticeably red, its ruddy cast contrasting well with hot, blue-white Rigel, the foot star at the opposite corner of the figure.

Betelgeuse is a relatively cool late-stage red giant, about 640 light years distant and a thousand times the diameter of the Sun. Such massive, dying stars may well become supernovae, blasting nearly all of their contents outwards in one final cataclysmic explosion and seeding the space around them with heavy elements cooked up in the hottest of cosmic ovens. If Betelgeuse does die in this spectacular way, it will become a fabulously bright star, although it's too far from Earth to do it any damage. Look at it often. You could be the first person since 1604 to see a supernova appear in our galaxy, the Milky Way—though it will of course have exploded 640 years ago. Light takes time to travel!

Deneb, Tail of the Swan

The stars can hold many surprises. For example, the nearest star to the Sun, Proxima, can be invisible, not only because it is more than 62° south of the celestial equator, making it difficult to spot at certain latitudes, but also because it is a dim red dwarf massively outshone by its neighbor Alpha Centauri. Yet one of the brightest of the summer stars, the blue-white giant Deneb, lay more than a thousand light years away! Deneb is the least bright of the three stars of the Summer Triangle (see Fig. 4.22). A good time to look at the Triangle is between August 11 and 13, when meteors emanating from the constellation of Perseus flash across it. See Chart 5 earlier in this section.

Chart 9 Fomalhaut finder chart

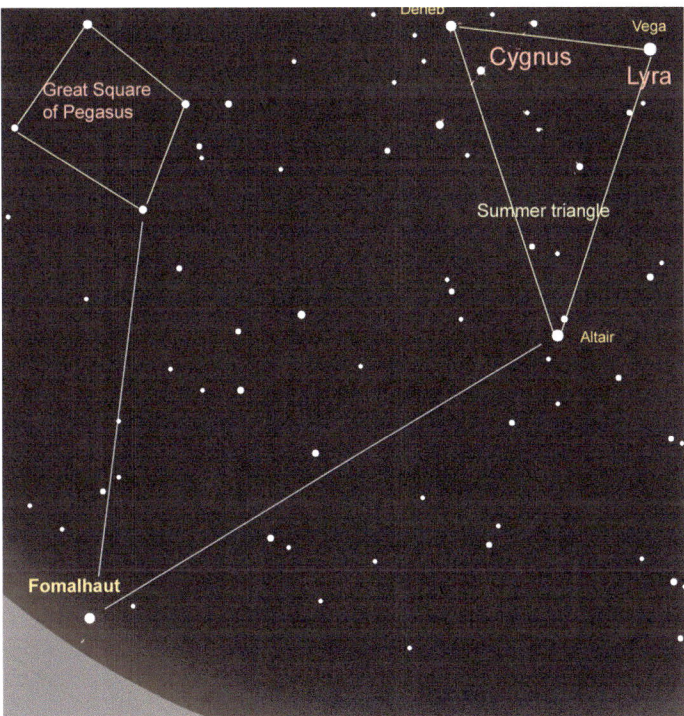

Fomalhaut, the Fish's Mouth

The seasonal chart for autumn (fall) shows Fomalhaut, a solitary bright star low in the southern sky during the late evening in September and October. Fomalhaut was seen by some ancient Arab astronomers as a celestial frog. It is in the constellation of the Southern Fish (Piscis Austrinus). Its name derives from Arabic *fum al-awt*, mouth of the fish, and its modern claim to fame is that, in November 2008, astronomers published a photo taken by the Hubble Space Telescope of an object that proved to be an extrasolar planet, orbiting just inside the ring of dusty debris around this star. See www.nasa.gov/mission_pages/hubble/science/fomalhaut.html.

Since then, both ground-based and space telescopes have discovered more than 2000 planets circling distant stars in the Sun's vicinity within the Milky Way. It's becoming obvious that they are not rarities. If that fact about stars and grains of sand on all the world's beaches is true, then how many planets must there be? And can ours really be the only planet where life began?

Chart 10 Garnet Star finder chart

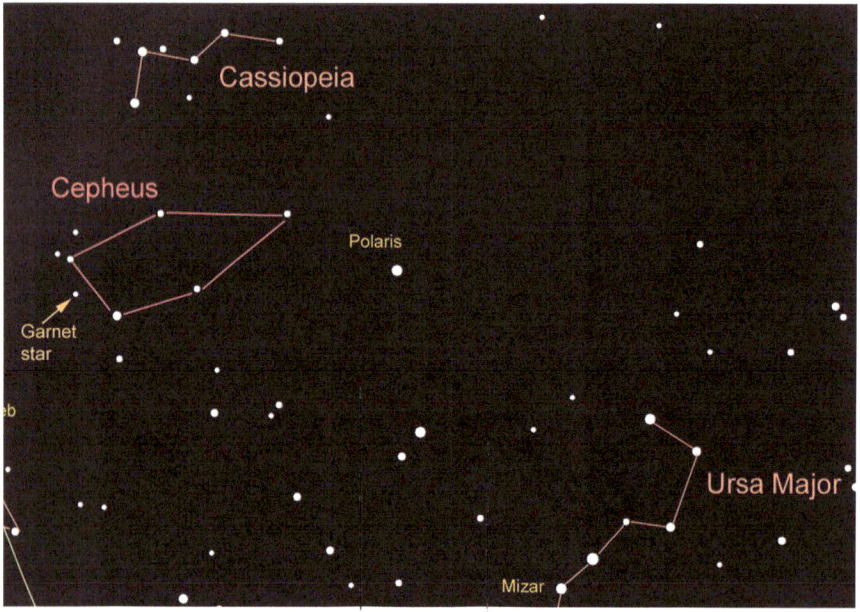

The Garnet Star, Herschel's Red Gem

At a glance, stars look like silver-white dots. A concentrated stare and a bit of comparison will suggest that they are not. Many show very subtle shades of red, yellow, blue-white…

There are several ancient and very red giant stars within easy reach of modern telescopes and binoculars, and a few are visible without optical aid. Some have telling names such as Hind's Crimson Star, La Superba, and the Garnet. The Garnet, an orange-red gem (see Fig. 4.24), shines in the circumpolar constellation of Cepheus, the King. It was named in the eighteenth century by Sir William Herschel, whose house in New King Street, Bath, southwest England, is now a superb museum (herschelmuseum.org.uk).

This reddest of stars is a pure joy to behold through a telescope. It is high above in summer in mid-latitudes, and blazes at a distance of about 6000 light years from us. The Garnet Star is 400,000 times brighter than the Sun.

Chart 11 Mizar finder chart

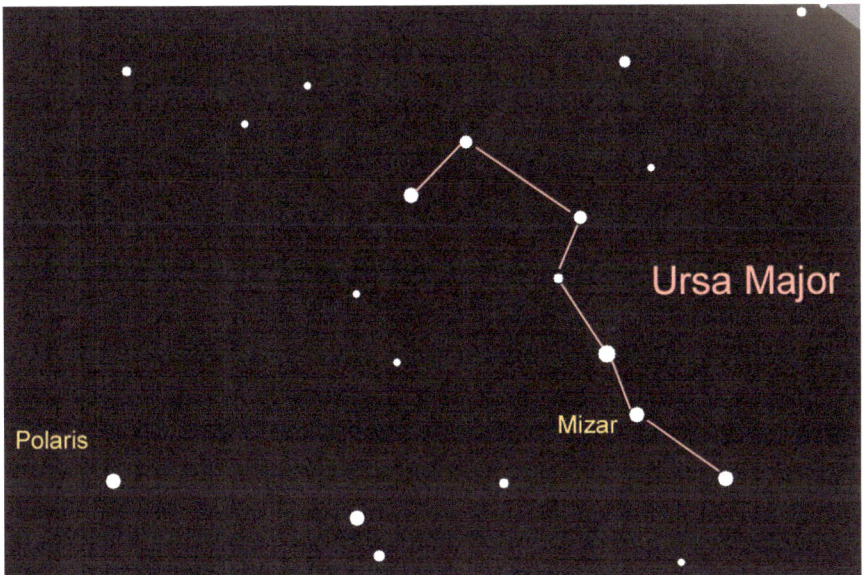

Mizar, the Second Hunter

The best time to go to a dark sky place might be December, when the glittering winter sky offers countless delights. At this time the seven brightest stars of the Great Bear, forming the Plough/Big Dipper, will not be far above the northern horizon (see Fig. 4.25). Native American legends held that leaves turned red in autumn as the Bear, wounded by arrows fired by the Hunters (the stars in the handle of the figure), bled upon them from its low position in the sky.

Look at the middle star of the Hunters. It's Mizar, and if your eyesight is good or your glasses efficient, you'll see a little star above and to the left of it. Binoculars will help locate it. This is Alcor, used as an eye test hundreds of years ago. The two

Fig. 4.24 The Garnet Star in Cepheus. (Photo: Simon Plumley)

Fig. 4.25 The Great Bear above the northern horizon (left). The bright star to the right is Capella in Auriga. (Photo: James Hilder)

stars were once called the Horse and Rider, and in mythology Alcor was also a cooking pot carried by the Second Hunter.

Mizar itself is a superb double star, and even a small telescope will split the pair—two white gems just 14 s of arc apart (for reference, the Moon is 1800 s of arc across).

In the telescope field there is a fainter star between Alcor and the Mizar pair. Interestingly, it's the faintest star with a name, and quite an impressive one, too: Sidus Ludovicianum (Ludwig's Star). It was so named by an eighteenth-century German astronomer, Liebknecht, who thought it was a new planet and called it after his sovereign, Count Ludwig of Hesse-Darmstadt.

Chart 12 Polaris finder chart

Polaris, the Pole Star

In Dave Strange's beautiful photo of an auroral glow in the sky over his home in Devon, southwest England (Fig. 4.26), you can follow the seven stars of the Plough/ Big Dipper downwards, and the last two, Merak and Dubhe, point to Polaris. Because it sits tight while all the other stars go around it all night long, it's a famous star—but it's not bright. Observers are often heard at star watches pointing to Venus, Vega or some other brilliant point of light and exclaiming "There's the Pole Star!" There are nearly 50 stars dotted around the sky that are brighter than humble Polaris. Its dogged stance in the north gave it long ago the name of Lode Star, lodestone being a naturally magnetized mineral used for north-pointing compasses.

Fig. 4.26 Auroral glow. (Photo: David Strange)

See James Castelli's intriguing video of circumpolar star motion filmed from Cherry Springs State Park, Pennsylvania, an International Dark-Sky Park, at www. youtube.com/watch?v=XTTDWhky9HY, or search "Stars circling Pole."

Polaris is about 400 light years away. Through a good-sized telescope a fainter companion star can be seen. Polaris hasn't always been at the north pole of the sky. Draco, the celestial Dragon's chief star Thuban was the north polar star 5000 years ago in the Age of the Pyramids, and brilliant Vega, in Lyra, will assume this role 12,000 years from now. Why? Because the axis of Earth's North Pole does not keep a fixed direction in space, but very slowly describes a huge circle, in the manner of an old-fashioned spinning top as it begins to fall over. This causes Earth's axis to point to different stars in the course of 26,000 years. The altitude of the Pole Star is the same as your latitude. So, from the far south of England, it is 50° high, and from Shetland, 60°. Observers in Florida will see the Pole Star between 25° and 31° high, and from the western section of the US/Canadian border the angle will be 49°.

Sirius, the Dog Star

Orion's Belt points downwards towards glittering Sirius (Fig. 4.27), the Sothis of the ancient Egyptians. They predicted the annual Nile floods by watching for the first glimpse of it in the dawn, rising before the Sun in the eastern midsummer sky.

Fig. 4.27 Orion's Belt points downwards toward Sirius, the brightest star in the night sky. (Photo: James Hilder)

Sirius (at 8.6 light years) is the beacon of the constellation of the Great Dog (Canis Major), faithfully following Orion through the sky. Interestingly, the Pawnee of North America refer to it as the Wolf Star.

Low down from temperate latitudes, it appears to sparkle vigorously in many colors, and this hot blue-white star is variously described in the historical literature as red, blue, green or yellow. Its companion, the 'Pup,' was first detected through an 18.5-inch (47-cm) aperture refracting telescope by American astronomer and telescope-maker Alvan Clark, from Northwestern University's Dearborn Observatory in January 1862. The observatory still organizes public viewing of the night sky. See Chart 8 earlier for how to find it.

The Moon: Our Nearest Neighbor—And Nature's Light Pollution

Our rocky satellite is about a quarter as wide as Earth—quite a big natural satellite given the size of the planet it orbits. Some astronomers think of the Earth-Moon system as a double planet. Any telescope or large mounted binoculars will give breathtaking images of the lunar surface. To seventeenth-century astronomers, peering through their imperfect "proportional glasses," our Moon seemed an Earthlike body. The dark patches of its giant impact basins and ancient basaltic flows were *maria* (Latin for 'seas'), and the lighter uplands and mountain ranges were continents (see Fig. 4.28).

Fig. 4.28 The full Moon. (Photo: Peter Kemp)

Fig. 4.29 The first-quarter Moon. (Photo: Peter Kemp)

However, it is the craters that really steal the show. These are seen at their best along the terminator, the line between daytime and night-time on the Moon, where shadows are most pronounced (see Fig. 4.29). Some are enormous. The crater Clavius, a scar left by a gigantic asteroidal impact about 4 billion years ago, is 140 miles (225 km) across and is peppered with smaller craters from more recent collisions. There are prominent craters such as Tycho, centered on spreading white rays of ejected material hundreds of miles long, craters with central peaks, 'ghost' craters half submerged in later lava flows; rilles (winding channels), scarps, isolated mountains, valleys… you can spend a whole evening looking at the Moon.

The Moon's brightness through the telescope can be fatiguing on the eye, though, and quite uncomfortable. Use a Moon filter, or wear sunglasses as an acceptable substitute. A full Moon has virtually no shadows, so craters look faint and far less interesting. The period of 2 weeks centered on the night of the full Moon is not the best time for starwatching, as its light dominates the sky and veils fainter objects.

Is the Moon really bigger when it's low down near the horizon? It can look huge, and seems so much smaller when up high. But it's just an optical illusion. Search 'Moon illusion' for a variety of explanations (which seems to suggest that nobody really knows why).

Ancient civilizations were intrigued by the mysterious motions of the five bright planets traveling across the starry background, sometimes reversing direction and then resuming their forward motion. With a little experience you can tell them apart: dull white Mercury, staying close to the Sun; Venus, so brilliant that it can be no other; Mars, the 'Red' Planet; Jupiter, usually bright and of a yellowish cast; and slow-moving Saturn, normally a less bright version of Jupiter. From a dark place, they will stand out in the southern sky (from north temperate latitudes), hardly twinkling, and inching nightly along the ecliptic, the path followed by the Sun throughout the year.

Mercury

Mercury (see Fig. 3.21) is an elusive planet. Because it's much nearer the Sun than we are, it's always seen somewhere in the general direction of our daystar, and stands not far from the horizon at dawn or dusk when visible at all. Its rapid motion around the Sun (a Mercurian year is 88 Earth days) caused the ancient Romans to identify it with their fleet-footed messenger of the gods. Consult planetarium programs such as *Night Vision* or search "observe Mercury" to find out when it can be seen, but NEVER look for it when the Sun is up. Mercury is small (3032 miles or 4879 km in diameter, not much larger than our Moon), so through a telescope it can be a bit disappointing, just a tiny gray disc with no visible surface features. Its heavily cratered terrain remained uncharted until *Mariner 10* reached it in 1974. It has the greatest temperature range in the whole of the Solar System. The planet rotates just once in 58 days, and during its months-long daytime it heats up to a broiling 427°C (800°F). At night the minimum temperature is −173°C (−280°F).

Venus

Venus (Fig. 4.30) is the brightest thing in the sky after the Sun and the Moon. It can cast distinct shadows in very dark places. Its planet-wide, highly reflective cloud envelope, composed mostly of sulfuric acid, makes Venus a mirror in the sky, bouncing sunlight into space. It is the dominant object among the far-off stars on moonless nights. The goddess of love's planet is almost the same size as Earth, but there the resemblance ends. The acidic clouds brew up Venus' very dense carbon dioxide atmosphere, raising it to a constant 470°C (878°F). The atmosphere is so dense that the pressure at the surface is 90 times Earth's air pressure. Venus is indeed Earth's terrible twin, a toxic and deadly place.

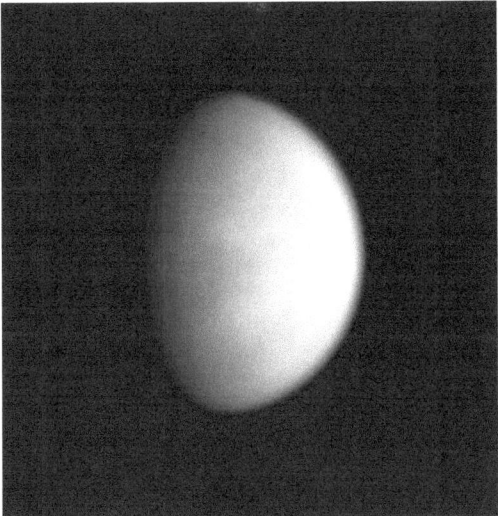

Fig. 4.30 The toxic clouds of Venus. (Photo: Bud Martin Budzynski)

Because, like Mercury, it is closer to the Sun than we are, it shows Moon-like phases through a well-adjusted telescope.

Mars

Mars (Fig. 4.31), more salmon-pink than red, is about half the diameter of Earth and a place of thin 'air' (carbon dioxide), wind-blown dust and big features. Its largest mountain, Olympus Mons, rises 16 miles (25 km, three times the height of Everest) above the surrounding boulder-strewn plains, and at its base Olympus is 374 miles (624 km) across. Mars' Mariner Valley complex is an enormous wind-scoured canyon system 2500 miles (4000 km) long. At its deepest, it dips 4 miles (7 km). It dwarfs the Grand Canyon in Arizona, which is 277 miles (446 km) long and just over a mile (1.6 km) deep. A medium-sized telescope will show a small pink disk, with possibly hints of dark surface markings and small highlights at the poles—the frozen carbon dioxide/water-ice caps.

Asteroids

These are not easy to find. Those easily visible through binoculars and telescopes (the best known being Ceres, Vesta, Pallas and Eros) look like faint stars. Indeed, the word asteroid means 'starlike.' Without expert advice and/or a detailed chart of an

Fig. 4.31 Mars, with features visible through large telescopes in perfect skies. (Photo: Bud Martin Budzynski)

asteroid's position among the stars, it's difficult to know which dot it is, if you can find it at all. Most asteroids orbit between Mars and Jupiter, and the largest is Ceres, 590 miles (950 km) in diameter, and upgraded in 2006 to a dwarf planet. Several asteroids have been photographed by space probes. A selection of images is available at space.about.com/od/picturesofasteroids/ig/Asteroids-Pictures-Gallery.

Jupiter

Jupiter (Fig. 4.32) is an enormous gas planet 1300 times the size of Earth. Through a large telescope it shows distinct bands in its atmosphere. You might be able to make out the Great Red Spot, a monster storm three times wider than Earth. It has been rolling around Jupiter for hundreds of years. A striking feature of the planet is that it looks slightly flattened at the poles; its rapid ten-hour rotation period makes the equatorial region bulge outwards. In 1610, with a rudimentary telescope, Galileo discovered the four brightest moons of Jupiter, which often appear strung out in a line, level with its equator. These Galilean satellites, mere specks of light through the eyepiece, have amazingly different terrains. You can explore them at en.wikipedia.org/wiki/Galilean_moons

Fig. 4.32 Jupiter. (Photo: Bud Martin Budzynski)

Saturn

Saturn (Fig. 4.33) is "the planet with the rings." Well, not quite. All four of the outer gaseous planets have rings, and they boast a sizable collection of moons, too. Even a small telescope will show Saturn's rings. Sometimes they are edge-on to the observer, and can't be seen at all. To date (early 2016), Saturn has 53 named moons. Titan, the biggest, has a dense methane and hydrocarbon atmosphere; Enceladus ejects plumes of ice-dust from its south polar region into the planet's ring system; Mimas has an enormous crater, making it look a little like the Death Star. Find the Cassini spacecraft website (www.nasa.gov/mission_pages/cassini) and marvel at what it has discovered since January 2005, when Cassini reached the Saturnian system and dropped a probe called Huygens onto the surface of Titan. This spacecraft has transmitted a mass of revelations, including stunning images of Saturn's icy rings and a range of intriguingly different satellites. The vast, tenuous ball of Saturn's globe would famously float in water if there were a tub big enough to accommodate it!

Fig. 4.33 Saturn. (Photo: Bud Martin Budzynski)

Uranus and Neptune

These are the two remaining gas giant planets, though nowadays they are increasingly called 'ice giants.' They can't be seen without optical aid, which means that they had to be discovered—Uranus in 1781 by William Herschel, from the garden of his house in New King Street, Bath, England (herschelmuseum.org.uk), and Neptune by Johann Galle from the Berlin Observatory in 1846. Through the telescope they are small, featureless bluish disks.

Pluto

Pluto was discovered by Clyde Tombaugh from the Lowell Observatory, Flagstaff, Arizona, in 1930. A speck in even the largest amateur telescopes, Pluto went from full planet status to officially becoming a dwarf planet in 2006. Try not to feel sorry for it, though. Pluto is no longer the little runt planet at the edge of the Solar System, but a big fish among the teeming shoals of rocky bodies, millions of them, that circle the Sun at enormous distances outside the orbit of Neptune. A remarkable feat of engineering delivered the New Horizons spacecraft on time to Pluto and its large moon Charon in July 2015, and photos of it are at www.nasa.gov/mission_pages/newhorizons/images.

New Horizons continues its voyage, on its way to encounters with more of the Solar System's rocky outliers.

Unpredictable and swift, meteors are arguably the most exciting things that happen at star watches. The short-lived streak of light marks the end of a particle of cosmic dust or debris, mostly between one mm and one cm in diameter, which may have been traveling through the Solar System for billions of years before colliding with our upper atmosphere at speeds ranging from 6 to 43 miles (10–70 km) a second (Fig. 4.34). The average meteor burns at an altitude of between 50 and 75 miles (80–120 km).

Particles entering the atmosphere are known as meteoroids, and, technically, the flash that we see is the meteor. If a larger space rock barrels its way through the air and hits Earth, it's called a meteorite.

Earth's (almost) circular orbit encounters streams of debris left by the slow disintegration of ancient comets. As they burn, they may appear very faint, so a dark observing site is by far the best place to be for the most meteors. Normally you may see a few 'shooting stars' (they're not stars at all, of course) in the course of an

Fig. 4.34 A meteor flashes below the star clusters of Taurus. (Photo: Chris Bowden)

evening's viewing, but on certain dates your chances are much increased as our planet meets those clouds of cosmic debris.

The most reliable meteor events, when they may fall at a rate of one every couple of minutes, or even more during exceptional surges, are the Lyrids, peaking on April 22, the Perseids (August 10–13), the Leonids (November 17–18) and the Geminids (December 13–14). Meteors are named after the constellations from which they appear to radiate into the sky. The point from which they appear to diverge is called the radiant. A bright Moon may spoil the show, so observing during the 2 weeks around new Moon, when the Moon is not far from the position of the Sun, is advisable.

When watching for meteors, stay comfortable! Craning your neck upwards for long periods is an effective way to convince yourself that you need to be doing something else. Lying down is the best way to hunt meteors. Bring a folding lawn lounger, and cover up with a blanket or two. Don't stare at the sky radiant from which the meteors appear to emanate—they could appear anywhere. Look around very slowly in all directions.

You might like to record what you've seen. You can draw the meteors' paths on a star chart (there are many websites offering these free, for example www.sky-maps.com), making a note of times, duration, brightness, color, and any associated phenomena such as trains of sparks or gas trails. Use a dim red light.

Photographing meteors is very much a hit-and-miss activity, as they usually seem to prefer rushing through an area of the sky your camera isn't pointed towards. Fix your camera to a steady mount aimed at an area of sky about 20–30° from the radiant.

Meteor-watching doesn't always have to be a 'low-tech' activity that uses no telescope or binoculars. Obviously, you are much more likely to notice the flashes if you are looking with the unaided eye at large areas of the sky rather than at the very restricted field of a telescope or binoculars. But occasional very bright meteors (the brightest are known as fireballs or bolides) leave persistent trails of ionized gases behind them. Perseids, Leonids and Geminids are particularly well known for this. The wakes of such very bright meteors can be studied, immediately after the flash, through binoculars. The gas trains linger and slowly change shape, glowing in the nighttime upper atmosphere against the black sky background. You may witness their slow dance, twisted by the motions of the upper atmosphere, then gradual fading to a pale gray 'snake,' falling to invisibility about 3 min after the initial flash.

So, with warm clothing, a red light and binoculars, hunt meteors. Don't forget your glasses if you need them for long distances, or in this case very long distances!

Artificial Satellites

Lying back on your lounger and contemplating the heavens, you are bound to see, after a while, what looks like a star slowly crossing the sky. If it's a single starlike point, and not flashing red and green lights, it's an artificial satellite. Many are visible to the unaided eye in a dark sky, monitoring our weather, aiding communications

Fig. 4.35 An Iridium satellite flares over Coylumbridge, Scottish Highlands. (Photo: Graeme Whipps)

and performing many other tasks. The Iridium series of satellites can be particularly bright, reflecting the Sun's light towards us (Fig. 4.35) as they slowly rotate. There are websites that tell you where to look for these flashes, known as Iridium flares (for example www.heavens-above.com/IridiumFlares.aspx). It's very easy to see the International Space Station, about four times brighter than Sirius. It rises in the west, and if passing high overhead takes about 5 min to cross the sky, eventually disappearing into Earth's shadow.

With some optical aid, thousands of intriguing stellar and non-stellar objects in the vicinity of the Sun are within reach. Specialized star atlases, planetarium programs and websites will reveal their whereabouts.

Star Clusters

Stars are normally born in groups, and there are large numbers of star clusters, both condensed and loose, that are visible through binoculars and telescopes. Some nearby open clusters are naked-eye treats: the hot, blue-white Pleiades (Fig. 4.36)

Fig. 4.36 The Pleiades and their associated nebulosity. (Photo: Shaun Reynolds)

are a fine example of a relatively young family of stars, while the Beehive Cluster (see Fig. 1.15 earlier) in Cancer is fainter, a mere smudge to our eyes but a beautiful sight with optical aid. Wonderful telescopic clusters include the Double Cluster in Perseus (at the top of Fig. 8.8 below) and M13 (see Fig. 4.21 above), a globular mass of a third of a million stars in Hercules.

Nebulae

Within our Milky Way Galaxy, countless clouds of gas glow due to irradiation by the stars associated with them. Dust is silhouetted against the light from stars or gas clouds behind it. The most striking nebula in the northern hemisphere of the sky is M42, the Orion Nebula (Fig. 4.37), looking like an out-of-focus star, and a joy through telescopes. Dark nebulae are harder to see, but a few such as the Horsehead Nebula, also in Orion, can be made out through large instruments. Planetary nebulae, caused by aging stars shrugging off their outer gaseous envelopes, appear as discs and sometimes ring-like structures. M57 (see Fig. 5.3 later), the Ring Nebula in Lyra, is the most accessible example.

Fig. 4.37 The convoluted gases of the Great Orion Nebula, with the Running Man Nebula to the left. (Photo: Iain Cartwright)

Supernova Remnant

Only one cloud of gas marking the explosive end of a massive star is easily accessible through a modest telescope: the faint Crab Nebula (M1) in Taurus, remnant of a supernova seen by oriental astronomers as a brilliant 'new' star in 1054. See apod. nasa.gov/apod/ap150816.html.

Faint Fuzzies: The Very Far Universe

The universe outside the Sun's realm is called the deep sky. The individual stars we see are relatively near, in our own part of the Milky Way Galaxy. More remote, telescopic deep-sky objects, distant galaxies of billions or trillions of stars, are faint and diffuse, with the exception of the Andromeda Galaxy (M31, our large neighbor galaxy in the Local Group) and the Magellanic Clouds, the Milky Way's largest satellite galaxies. To help imagine such remoteness, consider the light from M31. Traveling at 186,000 miles (300,000 km) per second, it takes just over 2.5 million years to reach us. The light from the most distant detectable galaxies left them 13 billion years ago, more than three times the age of Earth.

There are two main classes of galaxies: spiral and elliptical. Elliptical galaxies appear round or oval, and often have bright centers. Spiral galaxies, for example the

Fig. 4.38 Impressive detail in IC 443, the Jellyfish Nebula in Gemini. (Photo: Nik Szymanek)

Milky Way or M31, are disk-shaped with a central bulge, and if seen edge-on look like faint spindles in the sky. Using modern computer processing, amateur astronomers are now taking photos, many quite staggering in their detail, of deep-sky objects both within our own galaxy and beyond (Figs. 4.38 and 4.39).

Fig. 4.39 M104, the Sombrero Galaxy in Virgo, 28 million light years away. (Photo: Nik Szymanek)

Chapter 5

The Great American
Eclipse of August 2017
and Beyond

The total eclipse of 2017 in the United States is climbing up the news agenda. Hotels and cruises are booked, seats are reserved on airlines across the world, and state and county officials are already working on traffic management strategies as the big day ceases to be just an interesting footnote on the far horizon. In fact, it has become a huge national event requiring ever more urgent attention and planning. On Monday, August 21, 2017, the Moon's shadow, about a hundred km (just over 60 miles) wide, will cross America.

If you can get somewhere within the lunar shadow's long path across the United States on the big day, and the clouds stay away, what can you expect to see and experience?

On August 11, 1999, this author stood in a recently cut cornfield near Louvicamp, a tiny village in northwest France. With me in our group of eleven were two other members of the Wessex Astronomical Society (Dorset, UK) and our families. The long-awaited visit of the great end-of-the-century European total solar eclipse was imminent. On that day, the huge dark circle of the Moon's umbral shadow touched down on Earth just before 09.31 UT/GMT (or 05.31 EST) off the Canadian coast, about 300 km (186 miles) south of Nova Scotia. The umbra finally reached land on the Isles of Scilly off the southwest coast of England at 10:10 GMT (11.10 British Summer Time). One minute later, and the shadow was on its traverse of Cornwall and Devon in England's southwest peninsula. On the Channel island of Alderney observers were hoping for one and a half minutes of

© Springer International Publishing Switzerland 2016

B. Mizon, *Finding a Million-Star Hotel*, The Patrick Moore Practical Astronomy Series, DOI 10.1007/978-3-319-33855-2_5

totality. Sadly, it was a cloudy August day in southwest England, and indeed across most of the country. The late Sir Patrick Moore, Britain's premier astronomy popularizer, led the nation in bemoaning the plight of the UK's observers, gazing vainly upwards beneath those far too familiar Atlantic cloud streams. Some observers were able to catch a glimpse of the eclipsed Sun from the southernmost parts of the peninsula through breaks in the cloud, but several of the Wessex Astronomical Society members who had chartered a boat to observe from the Channel off the coast of Dorset were disappointed.

However, across the Channel, above Louvicamp, the morning's drizzle had stopped, and the clouds had begun to break up. As we made our way in growing excitement from our hotel in Aumale through country lanes to our chosen observing spot, the Sun appeared, much to our jubilation, in an ever-growing patch of blue sky. We set up a telescope to project a solar image onto a screen, and spread a large white sheet on the ground to catch the shadow bands, narrow alternating light and dark parallel lines that move intriguingly across light-colored surfaces shortly before and after a total solar eclipse. We donned the special eye patches we had remembered to bring, looking like a band of comic pirates to the amusement of some of the other dozens of expectant people who had found their way to this elevated viewpoint. We would see, with our one dark-adapted eye, the faint streamers and outer coronal envelope of the Sun as soon as it disappeared behind the Moon, while our neighbors' unadapted eyes would still need precious time to adjust.

At first, almost unnoticed by many, the slowly moving Moon began to eat into the Sun's disk; eclipse glasses on! The special filters of our eclipse glasses (don't try to use anything else to look at the Sun—your eyes are too precious) allowed us to follow the Moon's achingly slow progress across the disk. The air temperature dropped noticeably, slowly, by a few degrees, and the quality of the daylight changed. It gradually became sallow, fading to a kind of amber.

As the moment of totality approached, a visible wall of darkness could be seen looming, rushing in from the west. It looked like a vast oncoming rain-front, speeding up the valley towards our field. There were audible gasps from all around. And then the Sun was suddenly gone, switched off. In its place, the black lunar disk, at once sinister and majestic. Pure black, bounded by a thin burnished gold ring. Long, ghostly gray petals radiated from the disk—solar streamers of rarefied gases lancing through the corona, the outer atmosphere of our local star, now blotted out for two minutes and ten seconds. Because we had donned the eye patches, all this was immediately apparent. Around the horizon the darkness was a kind of purple mist, and for the brief period of totality, telescopes and cameras recorded the event. I made a quick sketch of the eclipsed sun using binoculars, making very sure that the ringer-timer I had set would warn me to stop looking half a minute before the Sun reappeared.

And then, all too soon, the emerging Sun flashed out from behind the Moon. Its light, pouring through valleys on the edge of the Moon, was broken up to form tiny pinpricks of radiance: 'Baily's Beads,' rapidly growing into glittering diamonds of light—the 'Diamond Ring' effect.

It was over. Breathlessness, general jubilation, popping corks. Americans high-fiving Swedes, Japanese hugging Spaniards, and even the traditionally reserved English dancing with anyone within reach.

Such gatherings as these will happen across the United States on August 21 in 2017. The shadow of the Moon will be touching down upon Earth in the north Pacific, halfway between Japan and the United States at 16.48 UT (= GMT). The lunar shadow will rush eastwards across the vast ocean at supersonic speed. It will make landfall on the continental United States at 10:15 a.m. Pacific Daylight Time (17.15 GMT), hurtling across the coast of Oregon at 3,900 km per hour (2,400 miles per hour) between Lincoln City and Newport. No doubt, thousands will be gathered to greet its onrush, weather permitting. Those who await the landfall of the eclipse on the Oregon coast should see the Sun blotted out, from the center line, for 1 minute and 59 seconds.

In a near replay of the trans-American total solar eclipse of June 8, 1918, the Moon's shadow will cross the continent diagonally. The width of the lunar shadow path will average 108 km (67 miles) as it dashes through Oregon, Idaho, Wyoming, Nebraska, Kansas, Missouri, Illinois, Kentucky, Tennessee, Georgia, North Carolina and, finally, South Carolina. For details of a published road atlas of the eclipse zone see www.mreclipse.com/pubs/Atlas2017.html.

The lunar shadow will finally leave the American landmass and move out into the Atlantic from the coast of South Carolina, near Charleston (the 1918 path crossed further to the south in Florida). The great shadow will lift off the surface of the Atlantic and back into space southwest of the Cape Verde Islands, near West Africa. Below is the Fred Espenak/NASA/Goddard Space Flight Center diagrammatic chart showing the circumstances of the eclipse. More details about the terms used in the diagram are given below.

Total Solar Eclipse of 2017 Aug 21

Ecliptic Conjunction = 18:31:19.6 TD   (= 18:30:11.2 UT)
Greatest Eclipse = 18:26:40.3 TD   (= 18:25:31.8 UT)

Eclipse Magnitude = 1.0306   Gamma = 0.4367

Saros Series = 145   Member = 22 of 77

Sun at Greatest Eclipse
(Geocentric Coordinates)

R.A. = 10h04m03.9s
Dec. = +11°51'43.0"
S.D. = 00°15'48.7"
H.P. = 00°00'08.7"

Moon at Greatest Eclipse
(Geocentric Coordinates)

R.A. = 10h04m30.6s
Dec. = +12°16'32.8"
S.D. = 00°16'03.4"
H.P. = 00°58'55.7"

External/Internal
Contacts of Penumbra

P1 = 15:46:51.5 UT
P2 = 18:11:57.2 UT
P3 = 18:39:24.9 UT
P4 = 21:04:23.5 UT

Constants & Ephemeris

ΔT =   68.4 s
k1 = 0.2725076
k2 = 0.2722810
Δb = 0.0"   Δl = 0.0"
Eph. = JPL DE405

Circumstances at Greatest Eclipse: 18:25:31.8 UT

Lat. = 36°58.0'N   Sun Alt. = 63.9°
Long. = 087°40.3'W   Sun Azm. = 197.9°
Path Width = 114.7 km   Duration = 02m40.1s

Circumstances at Greatest Duration: 18:21:49.2 UT

Lat. = 37°35'N   Sun Alt. = 63.8°
Long. = 089°07'W   Duration = 02m40.2s

External/Internal
Contacts of Umbra

U1 = 16:48:36.1 UT
U2 = 16:49:36.1 UT
U3 = 20:01:39.6 UT
U4 = 20:02:34.4 UT

Geocentric Libration
(Optical + Physical)

l =   4.64°
b = -0.57°
c = 21.90°

Brown Lun. No. = 1171

F. Espenak, NASA's GSFC
eclipse.gsfc.nasa.gov
2014 Feb 22

Will you be standing in the path of the speeding Moon shadow as it visits America in August 2017? Most Americans have never known such an opportunity. About 12 million people actually live within the totality path. The number of people living within a day's drive of the track of totality is around 200 million. Add to this the number who will come from further afield, from countries all across the planet; more humans will probably experience this eclipse than any other in history.

Remember, if you have had one eye covered for a while before totality, you will immediately be able to see faint features around the black lunar disk, such as the Sun's corona and associated streamers, and even large fiery prominences rising from the hidden limb (edge) of the Sun. The partial phases (when part of the Sun, even the narrowest 'crescent,' is still visible) must be viewed through specialized eclipse glasses. NEVER use these in conjunction with any optical instrument.

As the shadow courses across America, the totality time lengthens. At its maximum, over Carbondale, Illinois, you will have 2 minutes and 40 seconds to marvel at the spectacle. At the South Carolina coastline, the duration of totality will have dropped to 2 minutes and 34 seconds. Sizable cities within the path of darkness include Idaho Falls, Idaho; Casper, Wyoming; Lincoln, Nebraska; Kansas City, Missouri; Nashville, Tennessee; Charleston, South Carolina. Close to the total eclipse path, seeing a thin remaining sliver of sunlight, are Portland, Oregon; Boise, Idaho; Omaha, Nebraska; Springfield, Illinois; Chattanooga, Tennessee; Charlotte, North Carolina; and Atlanta and Augusta in Georgia. A longer list of places on or very near the eclipse track is given in Appendix 5.

The rest of North America, Greenland, Iceland, much of South America and the fringes of Western Europe will be inside the much wider penumbral shadow of the Moon. Here, observers will see only a partial eclipse of the Sun. In the UK, a small partial eclipse will be seen, peaking at around 20.00 BST. The closer you are to the path of totality, the more of the Sun will be obscured. Precautions are still absolutely essential; only proper eclipse glasses and viewers should be used. Smoked glass, sunglasses and the like are unsuitable. For safe indirect viewing, allow the Sun's light to pass through a small hole in a sheet of card, and study the resulting solar image on a light-colored screen.

From any one location on Earth, a total eclipse of the Sun might be seen once every 375 years approximately. During the 21st century there are 68 of them, an average of two every three years. The longest has already happened. On July 22, 2009, in southeast Asia, the Sun was blotted out for a remarkable 6 minutes and 39 seconds. Here are GMT (UK) times for the 2017 eclipse:

Start of partial eclipse (first contact): 16.56.05
Start of total eclipse (second contact): 18.24.11
Maximum eclipse: 18.25.32
End of total eclipse (third contact): 18.26.51
End of partial eclipse (fourth contact): 19.51.16
UK circumstances: astro.ukho.gov.uk/eclipse/0412017

There are many time zone converter websites to find equivalent U. S. times, for example:

wwp.greenwichmeantime.com/to/usa

Other useful websites for eclipse information are:

www.eclipse2017.org/2017/in_the_path.htm
www.exploratorium.edu/eclipse/how.html

Although most people need not travel far to see an eclipse of the Moon, as these are usually visible over a vast area of Earth, the narrow, limited zone of the Moon's shadow path during a total solar eclipse means that most hopeful 'eclipse chasers' normally have to go a great distance to see one.

The section below lists and briefly describes all the total and partial solar eclipses from February 15, 2018 (the next eclipse after the American event of August 2017), up to and including the Atlantic/European total solar eclipse of August 12, 2026. Remember to book hotels and cruises well in advance. Eclipse tourism is a growing phenomenon! The charts for the eclipse predictions are reproduced here with the permission of their creator Fred Espenak and the NASA Goddard Space Flight Center (GSFC). See the end of the section for an explanation of the terms used within the diagrams.

Future Eclipses of the Sun

It cannot be stressed often enough that the primary concern when viewing eclipses is eye safety, and the proper eclipse glasses/viewers should always be used when any part of the Sun's disk is visible.

Partial Solar Eclipse of 2018 Feb 15

Geocentric Conjunction = 20:15:02.2 UT   J.D. = 2458165.343776
Greatest Eclipse = 20:51:18.6 UT   J.D. = 2458165.368965

Eclipse Magnitude = 0.5986   Gamma = -1.2117

Saros Series = 150   Member = 17 of 71

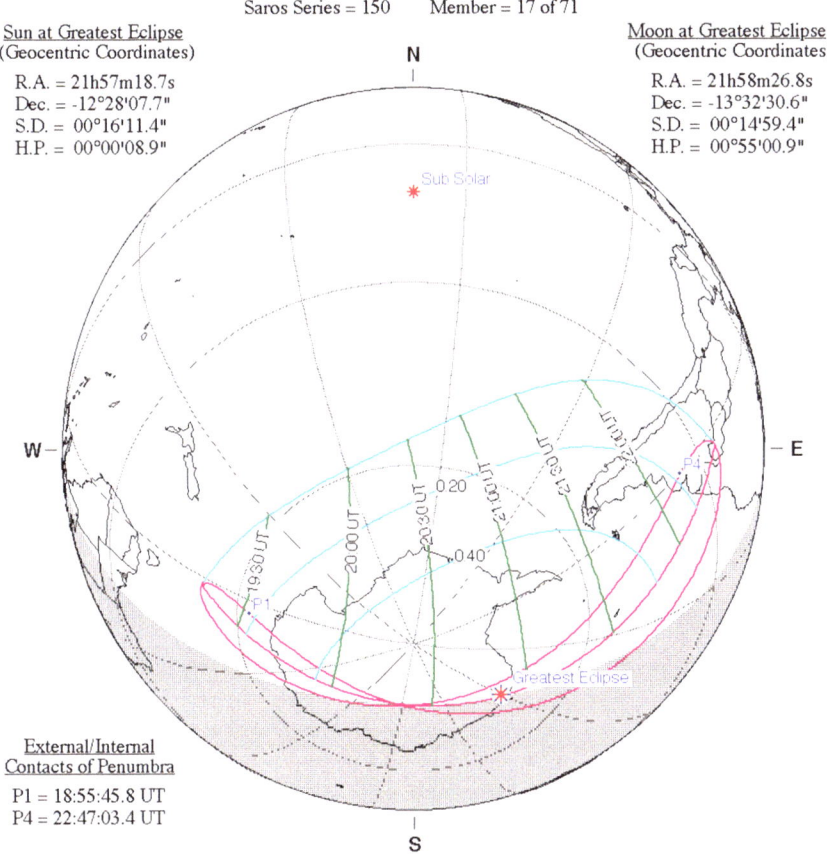

Sun at Greatest Eclipse
(Geocentric Coordinates)

R.A. = 21h57m18.7s
Dec. = -12°28'07.7"
S.D. = 00°16'11.4"
H.P. = 00°00'08.9"

Moon at Greatest Eclipse
(Geocentric Coordinates)

R.A. = 21h58m26.8s
Dec. = -13°32'30.6"
S.D. = 00°14'59.4"
H.P. = 00°55'00.9"

External/Internal
Contacts of Penumbra

P1 = 18:55:45.8 UT
P4 = 22:47:03.4 UT

Ephemeris & Constants

Eph. = Newcomb/ILE
ΔT = 74.8 s
k1 = 0.2724880
k2 = 0.2722810
Δb = 0.0"   Δl = 0.0"

Geocentric Libration
(Optical + Physical)

l = -3.95°
b = 1.41°
c = -21.58°

Brown Lun. No. = 1177

0   1000   2000   3000   4000   5000
Kilometers

F. Espenak, NASA's GSFC - Fri, Jul 2,
sunearth.gsfc.nasa.gov/eclipse/eclipse.html

The Antarctic partial solar eclipse of February 2018 will take place in the southern summer, the best of it visible (weather permitting) from rather inhospitable terrain where neither warm nor even temperate conditions can be expected. From Antarctica itself, the Sun will be nearly 49% covered by the Moon at maximum eclipse (see star symbol on map), and the solar disk will have been reduced to a broad U-shaped 'crescent'. As the eclipse diagram shows, the greatest coverage of the Sun will be visible, again weather permitting, from Queen Maud Land near the Greenwich Meridian, though the most practicable viewing will be from cruise ships in the Atlantic Ocean. Other land areas across which the Moon's shadow will be traveling on this day are the remote outcrops of South Georgia and the South Sandwich Islands, where a third of the Sun will be obscured, and the Falkland Islands, with the Sun only a quarter covered. South American countries catching the last of the event are Argentina, Chile, Uruguay and Brazil, all with decreasing Sun coverage by the Moon. In Brazil, at the end of the event, the Moon will obscure only a few percent of the solar disk.

Partial Solar Eclipse of 2018 Jul 13

Geocentric Conjunction = 03:08:59.5 UT　　J.D. = 2458312.631244
Greatest Eclipse = 03:01:02.4 UT　　J.D. = 2458312.625723

Eclipse Magnitude = 0.3367　　Gamma = -1.3541

Saros Series = 117　　Member = 69 of 71

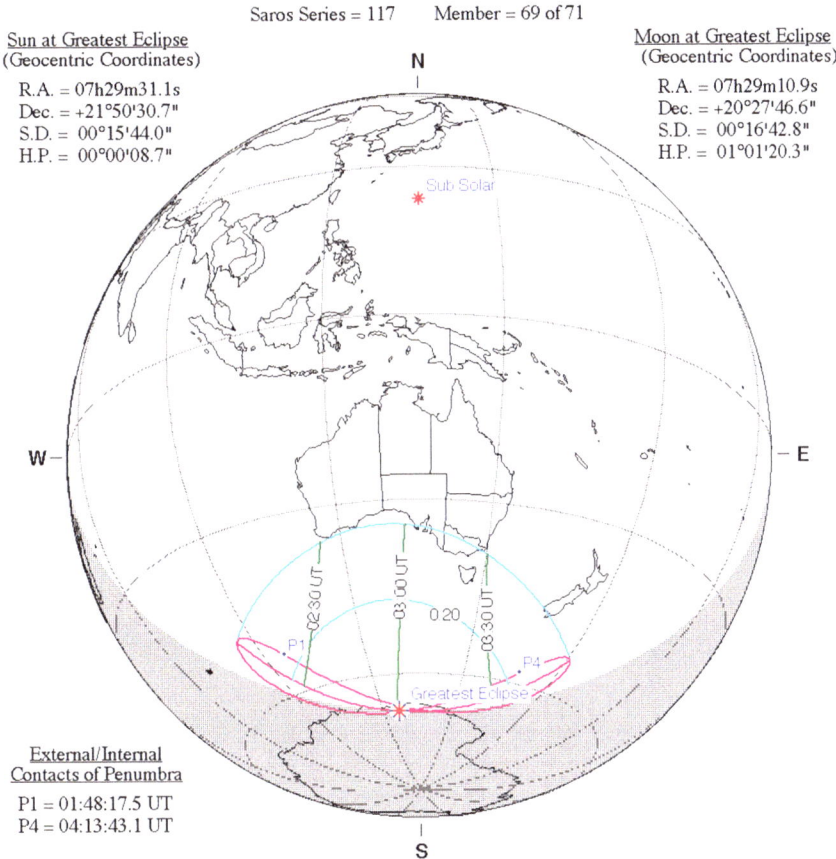

Sun at Greatest Eclipse
(Geocentric Coordinates)

R.A. = 07h29m31.1s
Dec. = +21°50'30.7"
S.D. = 00°15'44.0"
H.P. = 00°00'08.7"

Moon at Greatest Eclipse
(Geocentric Coordinates)

R.A. = 07h29m10.9s
Dec. = +20°27'46.6"
S.D. = 00°16'42.8"
H.P. = 01°01'20.3"

External/Internal
Contacts of Penumbra

P1 = 01:48:17.5 UT
P4 = 04:13:43.1 UT

Ephemeris & Constants

Eph. = Newcomb/ILE
ΔT = 75.2 s
k1 = 0.2724880
k2 = 0.2722810
Δb = 0.0"　Δl = 0.0"

Geocentric Libration
(Optical + Physical)

l = -0.38°
b = 1.79°
c = 10.14°

Brown Lun. No. = 1182

F. Espenak, NASA's GSFC - Fri, Jul 2,
sunearth.gsfc.nasa.gov/eclipse/eclipse.html

The second Antarctic partial solar eclipse of 2018 takes place in July, in the middle of the harsh far southern winter, and this time the greatest coverage of the Sun will be a little less than one quarter. This will be visible, weather permitting, from the coastal area of Antarctica due south of Australia. This is principally an Atlantic Ocean event, so observers may well again be predominantly on ships rather than on land, though partial eclipses do not draw as many eclipse chasers as do totals. Only a very small amount, around 5%, of the Sun, will be covered as seen from the very far south of Australia (the coastal area of South Australia and Victoria, and the island of Tasmania) and from the extreme far south of New Zealand.

Partial Solar Eclipse of 2019 Jan 06

Geocentric Conjunction = 01:43:34.7 UT   J.D. = 2458489.571929
Greatest Eclipse = 01:41:21.7 UT   J.D. = 2458489.570390

Eclipse Magnitude = 0.7147   Gamma = 1.1417

Saros Series = 122   Member = 58 of 70

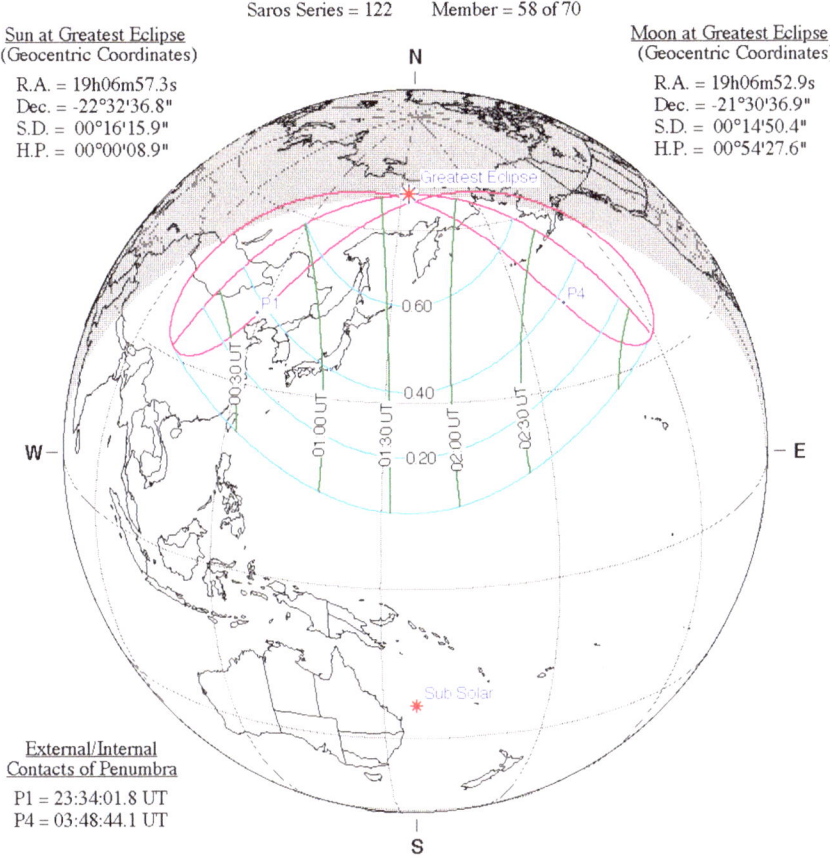

Sun at Greatest Eclipse
(Geocentric Coordinates)

R.A. = 19h06m57.3s
Dec. = -22°32'36.8"
S.D. = 00°16'15.9"
H.P. = 00°00'08.9"

Moon at Greatest Eclipse
(Geocentric Coordinates)

R.A. = 19h06m52.9s
Dec. = -21°30'36.9"
S.D. = 00°14'50.4"
H.P. = 00°54'27.6"

External/Internal
Contacts of Penumbra

P1 = 23:34:01.8 UT
P4 = 03:48:44.1 UT

Ephemeris & Constants

Eph. = Newcomb/ILE
ΔT =   75.7 s
k1 = 0.2724880
k2 = 0.2722810
Δb = 0.0" Δl = 0.0"

Geocentric Libration
(Optical + Physical)

l =   2.82°
b = -1.33°
c = -8.09°

Brown Lun. No. = 1188

0   1000   2000   3000   4000   5000
Kilometers

F. Espenak, NASA's GSFC - Fri, Jul 2,
sunearth.gsfc.nasa.gov/eclipse/eclipse.html

This first eclipse of 2019 will see the Moon's shadow (but not the dark umbral shadow of the total eclipse) move eastwards across a large area of northeast Asia, involving Japan, eastern China, Mongolia, North and South Korea and the far eastern regions of Russia, including the Kamchatka Peninsula. Much of the north Pacific Ocean, including Wake and Midway Islands, will also be in the shadow. Weather permitting, a small area of western Alaska and the Aleutians will see the final stages of the partial eclipse. At maximum eclipse the Sun will be about 50% covered, resembling a large bowl-like 'crescent' in the sky.

Total Solar Eclipse of 2019 Jul 02

Geocentric Conjunction = 19:21:36.4 UT   J.D. = 2458667.306672
Greatest Eclipse = 19:22:53.0 UT   J.D. = 2458667.307558

Eclipse Magnitude = 1.0459   Gamma = -0.6464

Saros Series = 127   Member = 58 of 82

Sun at Greatest Eclipse
(Geocentric Coordinates)

R.A. = 06h46m14.7s
Dec. = +23°00'36.5"
S.D. = 00°15'43.8"
H.P. = 00°00'08.6"

Moon at Greatest Eclipse
(Geocentric Coordinates)

R.A. = 06h46m17.8s
Dec. = +22°22'09.7"
S.D. = 00°16'14.9"
H.P. = 00°59'37.8"

External/Internal Contacts of Penumbra

P1 = 16:55:08.1 UT
P4 = 21:50:33.8 UT

External/Internal Contacts of Umbra

U1 = 18:01:04.3 UT
U2 = 18:03:24.6 UT
U3 = 20:42:19.6 UT
U4 = 20:44:44.3 UT

Local Circumstances at Greatest Eclipse

Lat. = 17°22.7'S   Sun Alt. = 49.6°
Long. = 108°58.8'W   Sun Azm. = 359.0°
Path Width = 200.6 km   Duration = 04m32.8s

Ephemeris & Constants

Eph. = Newcomb/ILE
ΔT =   76.2 s
k1 = 0.2724880
k2 = 0.2722810
Δb = 0.0" Δl = 0.0"

Geocentric Libration
(Optical + Physical)

l = -3.96°
b =   0.86°
c =   6.09°

Brown Lun. No. = 1194

0 1000 2000 3000 4000 5000
Kilometers

F. Espenak, NASA's GSFC · Fri, Jul 2,
sunearth.gsfc.nasa.gov/eclipse/eclipse.html

For the only fully total eclipse of the Sun in 2019, the Moon's shadow, 200 km (125 miles) wide, will begin its mostly ocean-bound traverse of Earth in the Pacific off New Zealand, touching down on the water just before 18.00 GMT before proceeding eastwards and passing not far to the north of the Pitcairn Islands. At a point on the track of totality in mid-ocean in the southern Pacific, the maximum duration of the eclipse will occur—4 minutes and 32 seconds. The shadow will make landfall on the west coast of Chile and cross the continent into Argentina, the event ending not far south of Buenos Aires at sunset, at 20.44 GMT. Here, as the Sun goes down, observers at the center of the shadow path will see it totally eclipsed for just over 2 minutes. La Serena in Chile and San Juan and Junín in Argentina would be good bases for viewing this eclipse. Interestingly for astronomers, two hours' drive to the north-northeast of La Serena is a famous astronomical site—the European Southern Observatory (ESO) complex at La Silla, strategically located at an altitude of 2,400 m beneath some of the darkest night skies on Earth on the outskirts of the dry plateau of the Atacama Desert. This ESO facility also lies in the path of totality for this eclipse, but it is quite a long way north of the lunar shadow's center line, and here, with clouds very unlikely, the duration of the total eclipse will be 1 minute 51 seconds, starting at 20.39 GMT. The associated partial eclipse will be visible across most of South America, with the exception of some northeastern coastal areas.

Annular Solar Eclipse of 2019 Dec 26

Geocentric Conjunction = 05:14:26.7 UT   J.D. = 2458843.718364
Greatest Eclipse = 05:17:36.0 UT   J.D. = 2458843.720556

Eclipse Magnitude = 0.9701   Gamma = 0.4135

Saros Series = 132   Member = 46 of 71

Sun at Greatest Eclipse
(Geocentric Coordinates)

R.A. = 18h17m56.6s
Dec. = -23°22'19.3"
S.D. = 00°16'15.7"
H.P. = 00°00'08.9"

Moon at Greatest Eclipse
(Geocentric Coordinates)

R.A. = 18h18m03.6s
Dec. = -22°58'50.6"
S.D. = 00°15'33.0"
H.P. = 00°57'04.1"

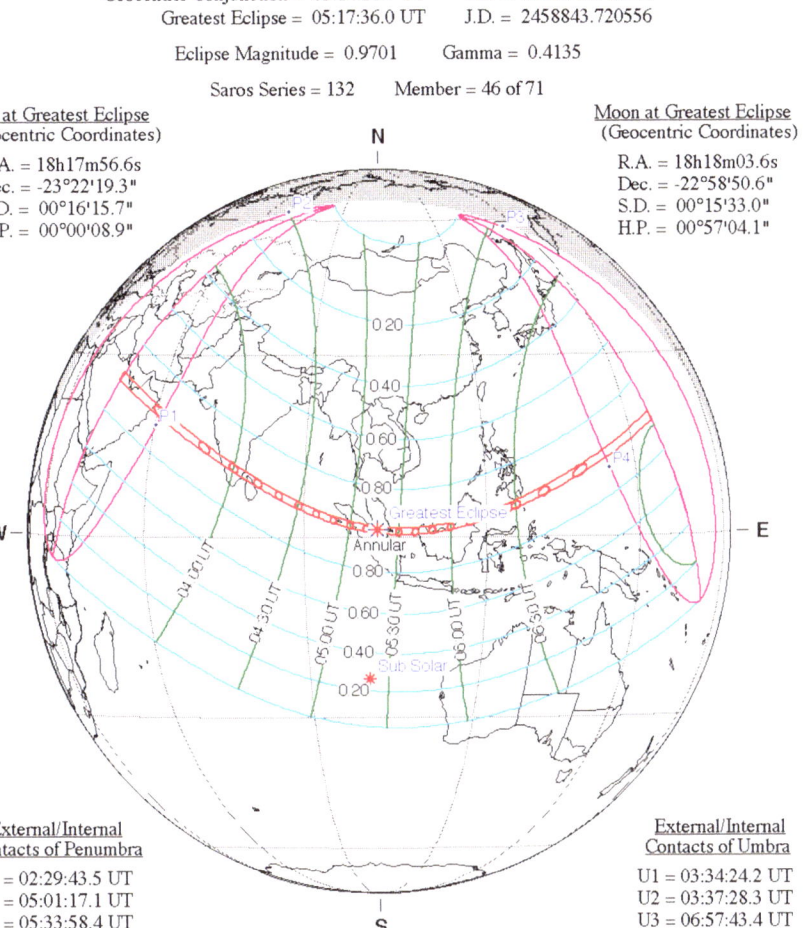

External/Internal Contacts of Penumbra

P1 = 02:29:43.5 UT
P2 = 05:01:17.1 UT
P3 = 05:33:58.4 UT
P4 = 08:05:36.1 UT

Ephemeris & Constants

Eph. = Newcomb/ILE
ΔT =   76.7 s
k1 = 0.2724880
k2 = 0.2722810
Δb = 0.0"   Δl = 0.0"

Local Circumstances at Greatest Eclipse

Lat. = 01°00.3'N   Sun Alt. = 65.6°
Long. = 102°16.5'E   Sun Azm. = 183.6°
Path Width = 117.9 km   Duration = 03m39.5s

External/Internal Contacts of Umbra

U1 = 03:34:24.2 UT
U2 = 03:37:28.3 UT
U3 = 06:57:43.4 UT
U4 = 07:00:53.6 UT

Geocentric Libration
(Optical + Physical)

l =   5.01°
b = -0.47°
c = -3.33°

Brown Lun. No. = 1200

0 1000 2000 3000 4000 5000
Kilometers

F. Espenak, NASA's GSFC - Fri, Jul 2,
sunearth.gsfc.nasa.gov/eclipse/eclipse.html

As 2019 draws towards its close, observers in the far northeast of Africa, a strip of southern India, northern Sri Lanka and many parts of Indonesia will experience one of the sky's most intriguing spectacles—an annular eclipse of the Sun. An annular eclipse (from the Latin *annulus*, a ring) occurs when the Moon, in its not exactly circular orbit around Earth, happens to be further away than average and cannot completely cover the solar disk. Because of their inequality in size, a ring of brilliant sunlight remains around the edge of the Moon when it centers itself on the Sun. The whole event will need to be observed through appropriate specialized solar filters, viewers or eyeglasses. Projection of the eclipse's image through a small hole onto a light colored screen is also pleasing—or you might use a colander or some other device with many small holes for an interesting multi-image effect. The radiating coronal features that become apparent during a total eclipse are not seen during annulars. The area from which the partial eclipse can be observed will be very large, stretching from the Arctic down to the far southern Atlantic.

Annular Solar Eclipse of 2020 Jun 21

Geocentric Conjunction = 06:41:18.4 UT   J.D. = 2459021.778685
Greatest Eclipse = 06:39:59.3 UT   J.D. = 2459021.777769

Eclipse Magnitude = 0.9940   Gamma = 0.1210

Saros Series = 137   Member = 36 of 70

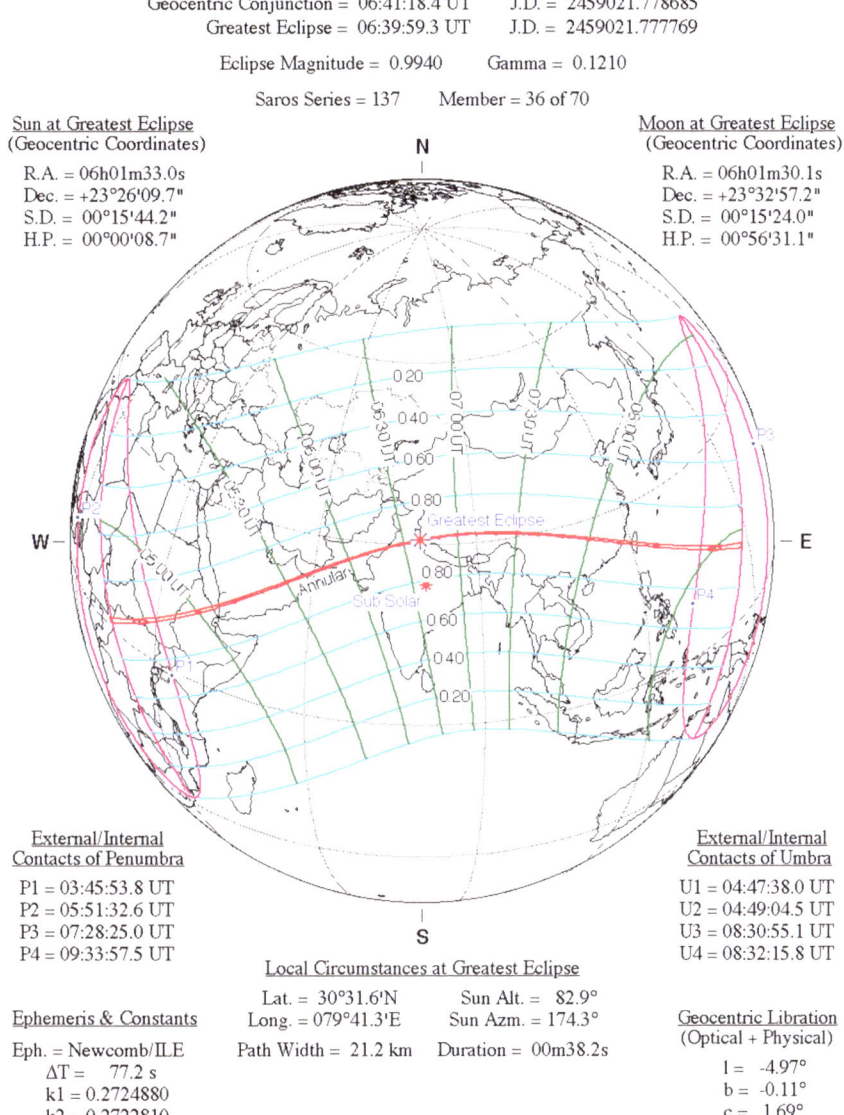

Sun at Greatest Eclipse
(Geocentric Coordinates)

R.A. = 06h01m33.0s
Dec. = +23°26'09.7"
S.D. = 00°15'44.2"
H.P. = 00°00'08.7"

Moon at Greatest Eclipse
(Geocentric Coordinates)

R.A. = 06h01m30.1s
Dec. = +23°32'57.2"
S.D. = 00°15'24.0"
H.P. = 00°56'31.1"

External/Internal Contacts of Penumbra

P1 = 03:45:53.8 UT
P2 = 05:51:32.6 UT
P3 = 07:28:25.0 UT
P4 = 09:33:57.5 UT

External/Internal Contacts of Umbra

U1 = 04:47:38.0 UT
U2 = 04:49:04.5 UT
U3 = 08:30:55.1 UT
U4 = 08:32:15.8 UT

Ephemeris & Constants

Eph. = Newcomb/ILE
ΔT =   77.2 s
k1 = 0.2724880
k2 = 0.2722810
Δb = 0.0"   Δl = 0.0"

Local Circumstances at Greatest Eclipse

Lat. = 30°31.6'N   Sun Alt. =   82.9°
Long. = 079°41.3'E   Sun Azm. = 174.3°
Path Width = 21.2 km   Duration = 00m38.2s

Geocentric Libration
(Optical + Physical)

l = -4.97°
b = -0.11°
c =   1.69°

Brown Lun. No. = 1206

0   1000   2000   3000   4000   5000
Kilometers

F. Espenak, NASA's GSFC - Fri, Jul 2,
sunearth.gsfc.nasa.gov/eclipse/eclipse.html

Another annular eclipse: the track this time will commence near the border between the two Congos (Congo Republic and Democratic Republic of Congo), and cross northeast Africa, the Arabian Peninsula, Pakistan, northern India (passing close to New Delhi), the Himalayas, southern China, Taiwan and ending in the western Pacific close to Guam. Note that this is a comparatively narrow eclipse path (at maximum only 21.2 km wide). Since the Moon will be covering over 99 % of the Sun, the greatest duration of this eclipse for observers can be only 38 seconds.

Total Solar Eclipse of 2020 Dec 14

Geocentric Conjunction = 16:18:05.4 UT   J.D. = 2459198.179230
Greatest Eclipse = 16:13:22.9 UT   J.D. = 2459198.175959

Eclipse Magnitude = 1.0254   Gamma = -0.2940

Saros Series = 142   Member = 23 of 72

Sun at Greatest Eclipse
(Geocentric Coordinates)

R.A. = 17h30m05.8s
Dec. = -23°15'32.2"
S.D. = 00°16'14.9"
H.P. = 00°00'08.9"

Moon at Greatest Eclipse
(Geocentric Coordinates)

R.A. = 17h29m54.2s
Dec. = -23°32'59.1"
S.D. = 00°16'23.7"
H.P. = 01°00'10.4"

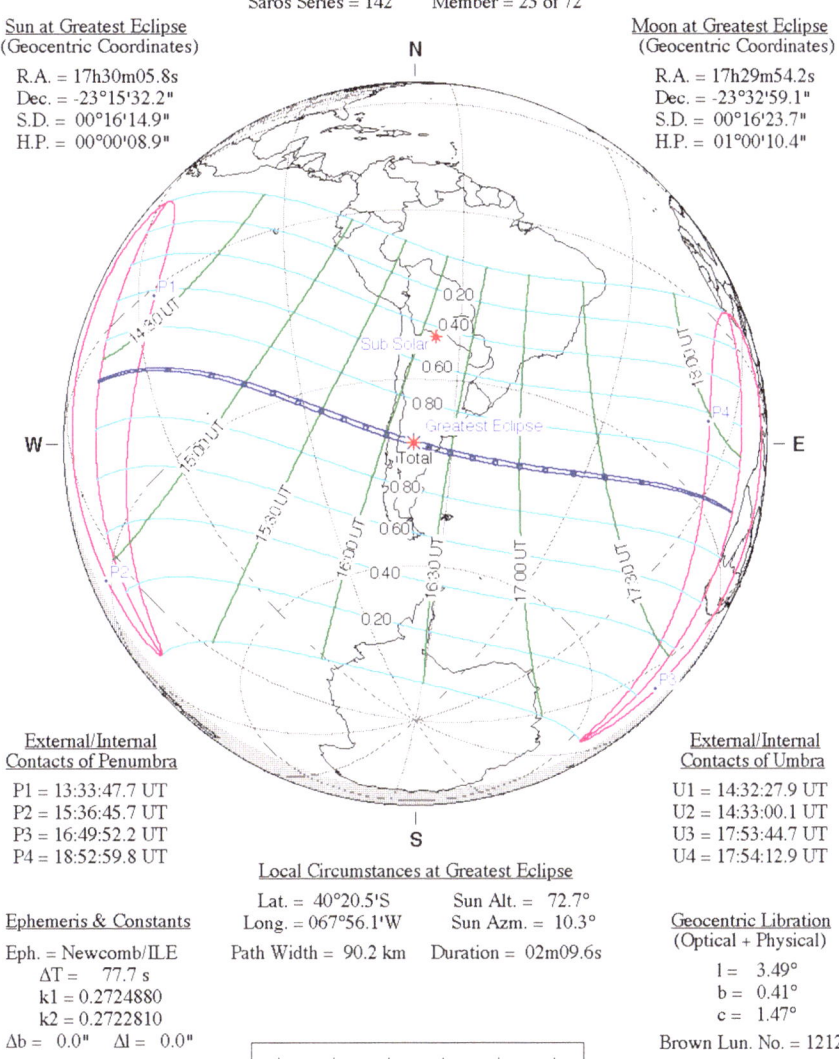

External/Internal
Contacts of Penumbra

P1 = 13:33:47.7 UT
P2 = 15:36:45.7 UT
P3 = 16:49:52.2 UT
P4 = 18:52:59.8 UT

Ephemeris & Constants

Eph. = Newcomb/ILE
  ΔT =   77.7 s
  k1 = 0.2724880
  k2 = 0.2722810
Δb = 0.0" Δl = 0.0"

Local Circumstances at Greatest Eclipse

Lat. = 40°20.5'S   Sun Alt. = 72.7°
Long. = 067°56.1'W   Sun Azm. = 10.3°
Path Width = 90.2 km   Duration = 02m09.6s

External/Internal
Contacts of Umbra

U1 = 14:32:27.9 UT
U2 = 14:33:00.1 UT
U3 = 17:53:44.7 UT
U4 = 17:54:12.9 UT

Geocentric Libration
(Optical + Physical)

l =   3.49°
b =   0.41°
c =   1.47°

Brown Lun. No. = 1212

0   1000   2000   3000   4000   5000
Kilometers

F. Espenak, NASA's GSFC - Fri, Jul 2,
sunearth.gsfc.nasa.gov/eclipse/eclipse.html

The second solar eclipse of 2020 will be visible from the Pacific or Atlantic oceans along most of the length of its track. The Moon's shadow will first meet the land on the small Chilean island of Isla Mocha just off Chile's Pacific coast. This is now the Mocha Island National Reserve, 48 square km/18.5 square miles in area. The shadow reaches the Chilean coast itself at 16.01 GMT (13.01 Chilean time), when the center line will be close to Puerto Saavedra, in Arauco Province, about 700 km/435miles south of the capital Santiago. Here, the shadow track is about 90 km/56 miles wide, and observers with clear skies on the center line will be treated to a total eclipse lasting for 2 minutes and 8 seconds. The duration of this eclipse is of similar length for the whole time that the shadow crosses South America. Towns in Chile in the path of the shadow include Temuco (at the northern edge), Gorbea and Villarrica. The shadow reaches the Argentinian border at 16.05 GMT, and thereafter passes over no large cities. The maximum eclipse of 2 minutes 9 seconds occurs around 16.13 GMT northwest of the small town of Sierra Colorada. It finally leaves the east coast of Argentina at about 16.25 GMT, near Viedma, and lifts off Earth, after a long journey across the Atlantic, about 600 km/373 miles west of Walvis Bay in Namibia.

Annular Solar Eclipse of 2021 Jun 10

Geocentric Conjunction = 11:00:58.7 UT   J.D. = 2459375.959013
Greatest Eclipse = 10:41:51.0 UT   J.D. = 2459375.945730

Eclipse Magnitude = 0.9435   Gamma = 0.9152

Saros Series = 147   Member = 23 of 80

Sun at Greatest Eclipse
(Geocentric Coordinates)

R.A. = 05h15m31.4s
Dec. = +23°02'37.1"
S.D. = 00°15'45.2"
H.P. = 00°00'08.7"

Moon at Greatest Eclipse
(Geocentric Coordinates)

R.A. = 05h14m53.5s
Dec. = +23°51'21.8"
S.D. = 00°14'46.8"
H.P. = 00°54'14.4"

External/Internal Contacts of Penumbra

P1 = 08:12:15.5 UT
P4 = 13:11:15.6 UT

External/Internal Contacts of Umbra

U1 = 09:49:43.4 UT
U2 = 10:00:36.9 UT
U3 = 11:22:53.4 UT
U4 = 11:33:44.7 UT

Local Circumstances at Greatest Eclipse

Lat. = 80°48.9'N   Sun Alt. = 23.3°
Long. = 066°48.3'W   Sun Azm. = 89.8°
Path Width = 527.1 km   Duration = 03m51.2s

Ephemeris & Constants

Eph. = Newcomb/ILE
ΔT =   78.2 s
k1 = 0.2724880
k2 = 0.2722810
Δb =   0.0"   Δl =   0.0"

Geocentric Libration
(Optical + Physical)

l =   -2.30°
b = -1.06°
c = -2.93°

Brown Lun. No. = 1218

0   1000   2000   3000   4000   5000
Kilometers

F. Espenak, NASA's GSFC - Fri, Jul 2,
sunearth.gsfc.nasa.gov/eclipse/eclipse.html

The first solar eclipse of 2021 crosses the Arctic. It will be annular, and will follow an unusual curving path, a phenomenon possible only in regions near the poles. This will not be the normal west-east track, but one first moving to the northeast across eastern Canada and Hudson Bay, then turning north and westwards over Baffin Island on its way across the North Pole and finally descending southwards across the East Siberian Sea into far eastern Russia. The track ends near the Siberian Kolyma River. In the northeastern United States and Canada, the Sun will rise already partially eclipsed. The fact that the path of annularity passes over the North Pole is itself of interest, and the last eclipse there will have been that of March 2015, when the total eclipse path which had crossed the Faroes and Svalbard ended at the pole at the spring equinox, as the Sun reappeared after the months-long polar night.

Total Solar Eclipse of 2021 Dec 04

Geocentric Conjunction = 07:56:04.9 UT   J.D. = 2459552.830612
Greatest Eclipse = 07:33:22.5 UT   J.D. = 2459552.814844

Eclipse Magnitude = 1.0367   Gamma = -0.9526

Saros Series = 152   Member = 13 of 70

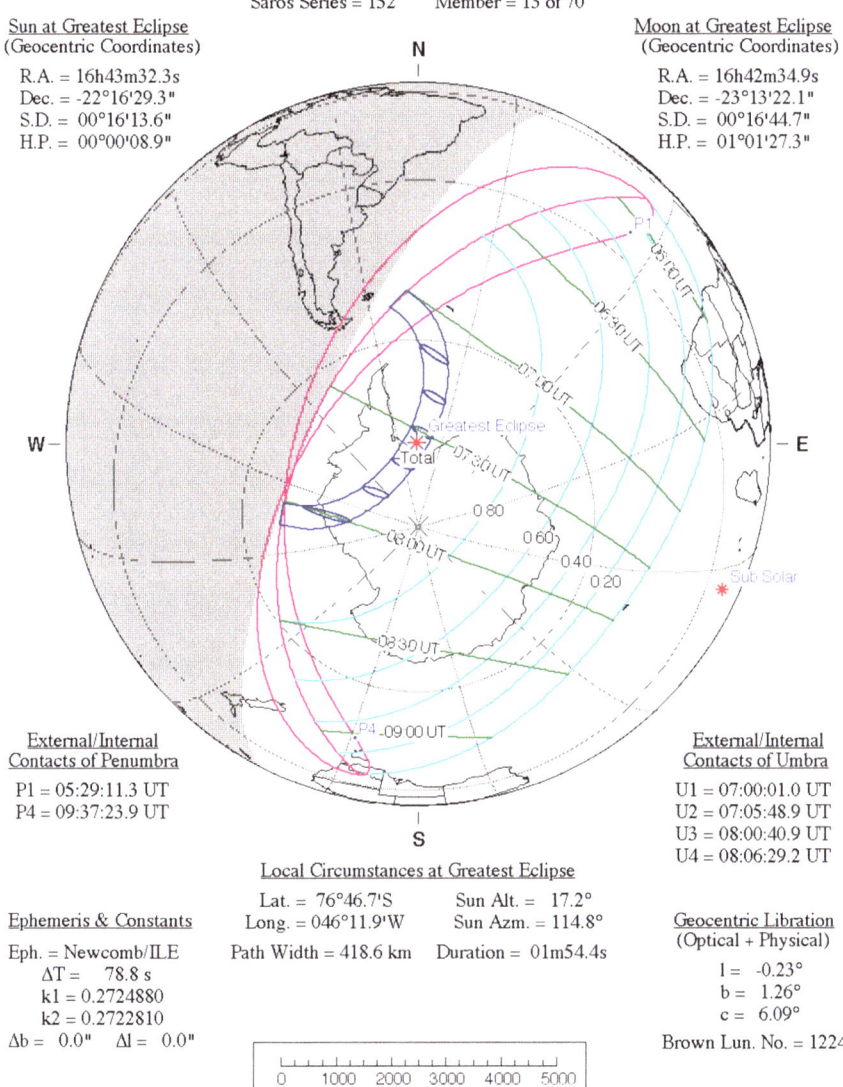

Sun at Greatest Eclipse
(Geocentric Coordinates)

R.A. = 16h43m32.3s
Dec. = -22°16'29.3"
S.D. = 00°16'13.6"
H.P. = 00°00'08.9"

Moon at Greatest Eclipse
(Geocentric Coordinates)

R.A. = 16h42m34.9s
Dec. = -23°13'22.1"
S.D. = 00°16'44.7"
H.P. = 01°01'27.3"

External/Internal
Contacts of Penumbra

P1 = 05:29:11.3 UT
P4 = 09:37:23.9 UT

External/Internal
Contacts of Umbra

U1 = 07:00:01.0 UT
U2 = 07:05:48.9 UT
U3 = 08:00:40.9 UT
U4 = 08:06:29.2 UT

Local Circumstances at Greatest Eclipse

Lat. = 76°46.7'S   Sun Alt. = 17.2°
Long. = 046°11.9'W   Sun Azm. = 114.8°
Path Width = 418.6 km   Duration = 01m54.4s

Ephemeris & Constants

Eph. = Newcomb/ILE
ΔT = 78.8 s
k1 = 0.2724880
k2 = 0.2722810
Δb = 0.0"   Δl = 0.0"

Geocentric Libration
(Optical + Physical)

l = -0.23°
b = 1.26°
c = 6.09°

Brown Lun. No. = 1224

F. Espenak, NASA's GSFC - Fri, Jul 2,
sunearth.gsfc.nasa.gov/eclipse/eclipse.html

This Antarctic eclipse will be something of a mirror image of the previous (Arctic) event earlier that year, but this time it will be total, with a wide curving path starting near South Georgia and the South Sandwich Islands, 2,700 km/1,680 miles east of Argentina. From these remote outposts the track will sweep down to curve around the Antarctic peninsula, the greatest duration of the event occurring close to the Antarctic coast and back into the Atlantic Ocean. Definitely one for the ship-borne observer, and it is very likely that many more penguins than people will be seeing this one!

Partial Solar Eclipse of 2022 Apr 30

Geocentric Conjunction = 19:40:42.5 UT   J.D. = 2459700.319937
Greatest Eclipse = 20:41:20.2 UT   J.D. = 2459700.362039

Eclipse Magnitude = 0.6389   Gamma = -1.1900

Saros Series = 119   Member = 66 of 71

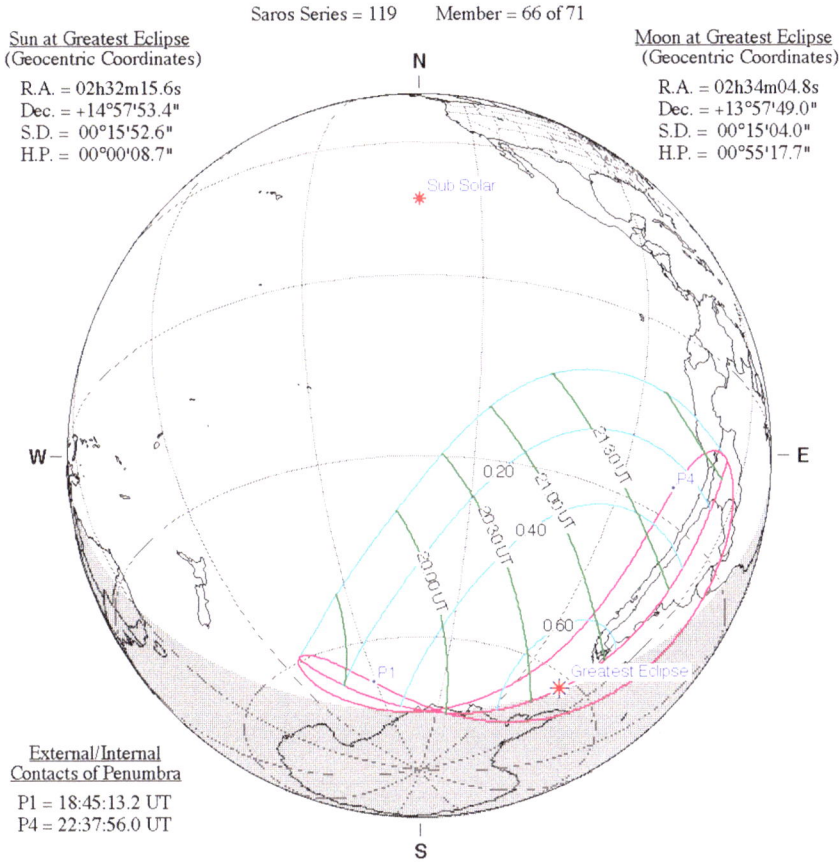

Sun at Greatest Eclipse
(Geocentric Coordinates)

R.A. = 02h32m15.6s
Dec. = +14°57'53.4"
S.D. = 00°15'52.6"
H.P. = 00°00'08.7"

Moon at Greatest Eclipse
(Geocentric Coordinates)

R.A. = 02h34m04.8s
Dec. = +13°57'49.0"
S.D. = 00°15'04.0"
H.P. = 00°55'17.7"

External/Internal
Contacts of Penumbra

P1 = 18:45:13.2 UT
P4 = 22:37:56.0 UT

Ephemeris & Constants

Eph. = Newcomb/ILE
ΔT =   79.2 s
k1 = 0.2724880
k2 = 0.2722810
Δb = 0.0"   Δl = 0.0"

Geocentric Libration
(Optical + Physical)

l =   4.01°
b = 1.40°
c = -16.62°

Brown Lun. No. = 1229

0   1000   2000 3000   4000   5000
Kilometers

F. Espenak, NASA's GSFC - Fri, Jul 2,
sunearth.gsfc.nasa.gov/eclipse/eclipse.html

Another southern event, this one a partial eclipse visible from a very small strip of the Antarctic coast, from the Atlantic Ocean, the South Pacific and much of the southern and western parts of South America. Unlikely to be seen by many.

Partial Solar Eclipse of 2022 Oct 25

Geocentric Conjunction = 10:03:36.7 UT   J.D. = 2459877.919175
Greatest Eclipse = 11:00:00.4 UT   J.D. = 2459877.958338

Eclipse Magnitude = 0.8611   Gamma = 1.0700

Saros Series = 124   Member = 55 of 73

Sun at Greatest Eclipse
(Geocentric Coordinates)

R.A. = 13h59m20.4s
Dec. = -12°10'16.6"
S.D. = 00°16'05.0"
H.P. = 00°00'08.8"

Moon at Greatest Eclipse
(Geocentric Coordinates)

R.A. = 14h01m10.8s
Dec. = -11°14'16.1"
S.D. = 00°15'52.6"
H.P. = 00°58'16.0"

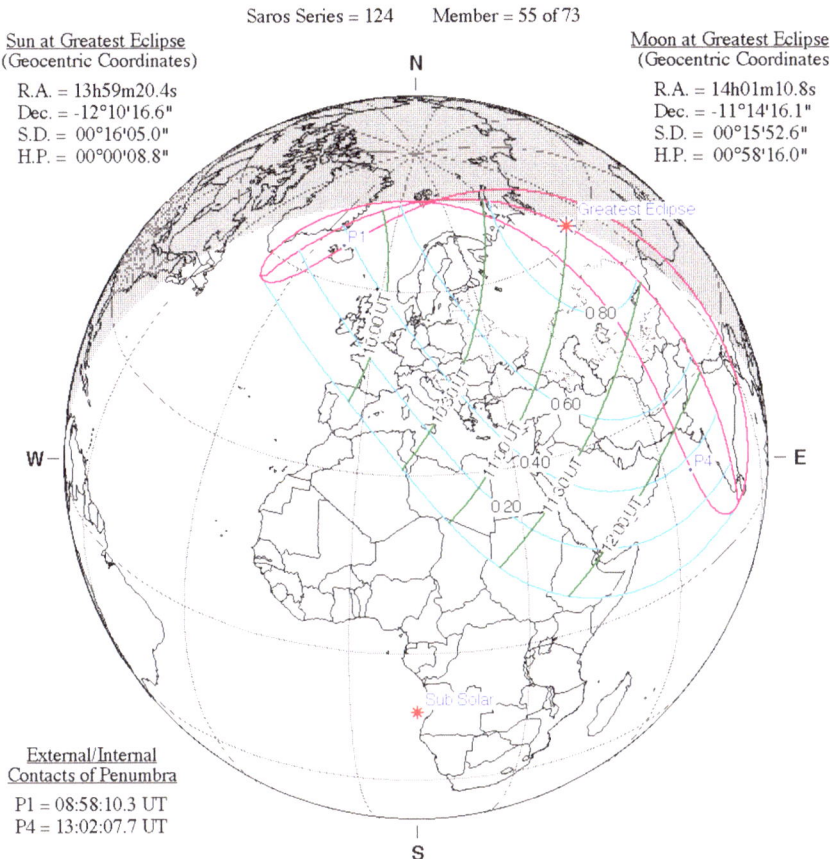

External/Internal
Contacts of Penumbra

P1 = 08:58:10.3 UT
P4 = 13:02:07.7 UT

Ephemeris & Constants

Eph. = Newcomb/ILE
ΔT =   79.7 s
k1 = 0.2724880
k2 = 0.2722810
Δb = 0.0"   Δl = 0.0"

Geocentric Libration
(Optical + Physical)

l =   -4.55°
b = -1.38°
c = 18.60°

Brown Lun. No. = 1235

| 0 | 1000 | 2000 | 3000 | 4000 | 5000 |
Kilometers

F. Espenak, NASA's GSFC - Fri, Jul 2,
sunearth.gsfc.nasa.gov/eclipse/eclipse.html

This partial eclipse, unlike the previous (Antarctic) event of 2022, should be seen, weather permitting, mostly over land, and by vast numbers of people in Europe, the Middle East, North Africa and western Asia, including most of India. At maximum eclipse, the Sun will appear as a slender 'crescent' (the word crescent really means 'growing,' but it is commonly used for such a shape even if it is shrinking—actually de-crescent).

Hybrid Solar Eclipse of 2023 Apr 20

Geocentric Conjunction = 03:55:26.5 UT   J.D. = 2460054.663502
Greatest Eclipse = 04:16:37.5 UT   J.D. = 2460054.678212

Eclipse Magnitude = 1.0132   Gamma = -0.3951

Saros Series = 129   Member = 52 of 80

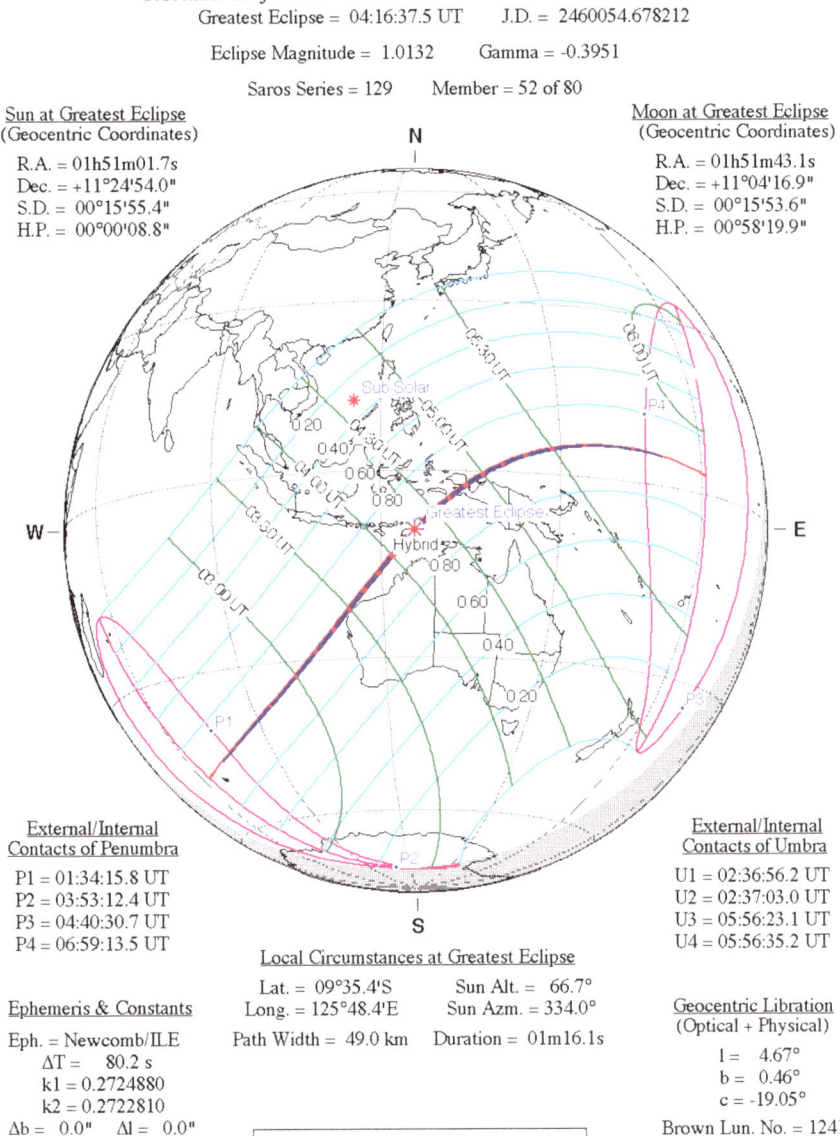

Sun at Greatest Eclipse
(Geocentric Coordinates)

R.A. = 01h51m01.7s
Dec. = +11°24'54.0"
S.D. = 00°15'55.4"
H.P. = 00°00'08.8"

Moon at Greatest Eclipse
(Geocentric Coordinates)

R.A. = 01h51m43.1s
Dec. = +11°04'16.9"
S.D. = 00°15'53.6"
H.P. = 00°58'19.9"

External/Internal Contacts of Penumbra

P1 = 01:34:15.8 UT
P2 = 03:53:12.4 UT
P3 = 04:40:30.7 UT
P4 = 06:59:13.5 UT

External/Internal Contacts of Umbra

U1 = 02:36:56.2 UT
U2 = 02:37:03.0 UT
U3 = 05:56:23.1 UT
U4 = 05:56:35.2 UT

Ephemeris & Constants

Eph. = Newcomb/ILE
ΔT =   80.2 s
k1 = 0.2724880
k2 = 0.2722810
Δb =   0.0"   Δl =   0.0"

Local Circumstances at Greatest Eclipse

Lat. = 09°35.4'S   Sun Alt. =   66.7°
Long. = 125°48.4'E   Sun Azm. = 334.0°
Path Width = 49.0 km   Duration = 01m16.1s

Geocentric Libration
(Optical + Physical)

l =   4.67°
b =   0.46°
c = -19.05°

Brown Lun. No. = 1241

```
|  |  |  |  |  |  |  |  |  |  |  |  |  |
0      1000    2000   3000    4000    5000
              Kilometers
```

F. Espenak, NASA's GSFC - Fri, Jul 2,
sunearth.gsfc.nasa.gov/eclipse/eclipse.html

This will be a hybrid eclipse, with the first and last portions of its narrow path near sunrise and sunset seen as annular, and the rest as total. Hybrid eclipses always start as annular, then change to total eclipses, and finally change back to annular. The eclipse of April 20, 2023, begins to the west of the very remote French-administered Kerguelen Islands, in the far south of the Indian Ocean, becoming total before clipping the edge of Australia at the Cape Range National Park near Exmouth in the state of Western Australia. Its next landfalls will be on the Muiron Islands and fish-shaped Serrurier Island off the Australian coast. Interestingly, Bessieres Island, not far north of Serrurier, will see both this eclipse and the total eclipse of December 2038. The track reaches Barrow Island at 03.34 GMT (where the duration of totality will be 1 minute and 5 seconds). After crossing the uninhabited Ashmore Reef (Pulau Pasir), the track reaches the eastern end of East Timor at 04.18 GMT, with Viqueque and Lospalos experiencing a total eclipse of 1 minute and 16 seconds. The last land area on the eclipse path will be in West Papua, whence the shadow path proceeds to its extinction in the middle of the South Pacific. The associated partial eclipse will be seen over a large area, including part of Antarctica, the Atlantic Ocean, the southern part of the Indian Ocean, south and east Asia (mostly Indonesia), Australia, New Zealand (North Island) and much of the southern and western Pacific.

Annular Solar Eclipse of 2023 Oct 14

Geocentric Conjunction = 17:36:28.8 UT   J.D. = 2460232.233667
Greatest Eclipse = 17:59:21.0 UT   J.D. = 2460232.249549

Eclipse Magnitude = 0.9520   Gamma = 0.3752

Saros Series = 134   Member = 44 of 71

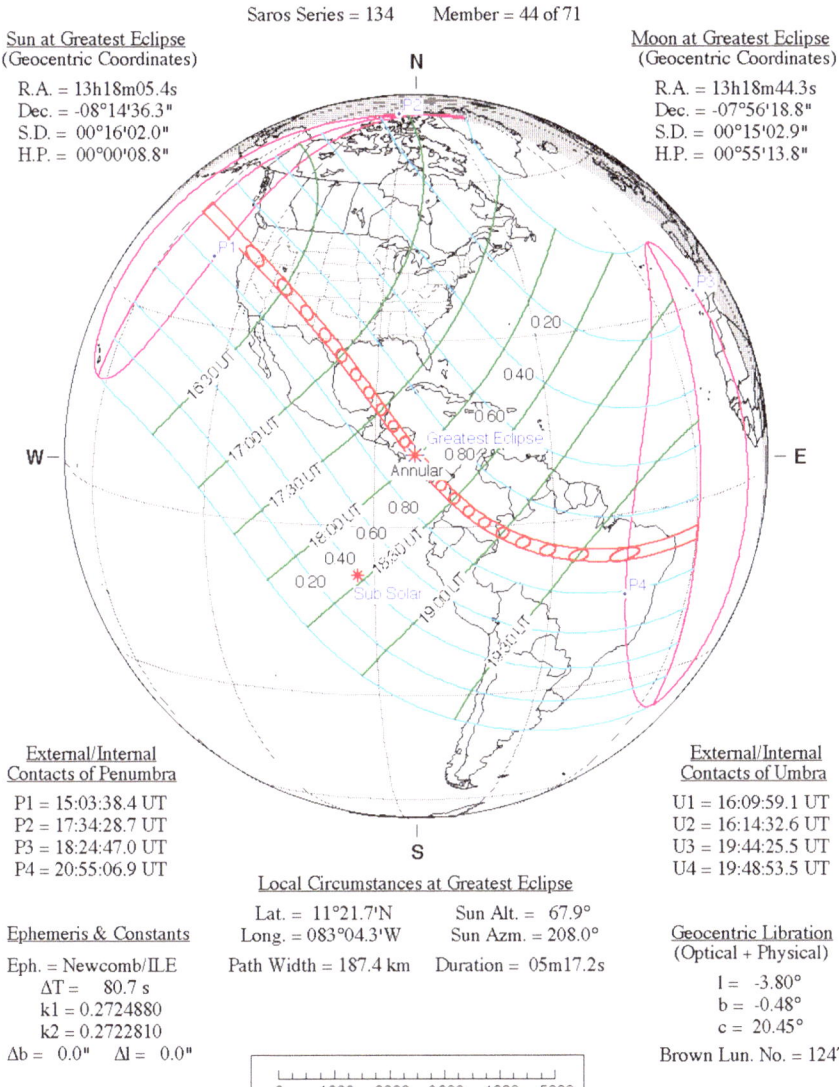

Sun at Greatest Eclipse
(Geocentric Coordinates)

R.A. = 13h18m05.4s
Dec. = -08°14'36.3"
S.D. = 00°16'02.0"
H.P. = 00°00'08.8"

Moon at Greatest Eclipse
(Geocentric Coordinates)

R.A. = 13h18m44.3s
Dec. = -07°56'18.8"
S.D. = 00°15'02.9"
H.P. = 00°55'13.8"

External/Internal
Contacts of Penumbra

P1 = 15:03:38.4 UT
P2 = 17:34:28.7 UT
P3 = 18:24:47.0 UT
P4 = 20:55:06.9 UT

External/Internal
Contacts of Umbra

U1 = 16:09:59.1 UT
U2 = 16:14:32.6 UT
U3 = 19:44:25.5 UT
U4 = 19:48:53.5 UT

Local Circumstances at Greatest Eclipse

Lat. = 11°21.7'N   Sun Alt. = 67.9°
Long. = 083°04.3'W   Sun Azm. = 208.0°
Path Width = 187.4 km   Duration = 05m17.2s

Ephemeris & Constants

Eph. = Newcomb/ILE
ΔT = 80.7 s
k1 = 0.2724880
k2 = 0.2722810
Δb = 0.0"   Δl = 0.0"

Geocentric Libration
(Optical + Physical)

l = -3.80°
b = -0.48°
c = 20.45°

Brown Lun. No. = 1247

0   1000   2000   3000   4000   5000
Kilometers

F. Espenak, NASA's GSFC - Fri, Jul 2,
sunearth.gsfc.nasa.gov/eclipse/eclipse.html

The Pan-American Eclipse of October 14, 2023, will be annular, its track running down through the United States, Mexico, Central America, Colombia and Brazil. Landfall occurs in Oregon at 16:18 GMT (08.18 PDT), very close to the point where the 2017 eclipse center line passed. It will cross the northeast corner of California and then move through Nevada, Utah, Arizona, New Mexico and Texas. Around 16.37 GMT, Albuquerque (New Mexico) will be on the center line, where the event will last for 4 minutes and 52 seconds. San Antonio and Corpus Christi (both in Texas) will be similarly placed. The eclipse will reach the Gulf of Mexico at Corpus Christi at 16.58 GMT. The annular path will then cross the Yucatan peninsula north of the town of Campeche at 17.24 GMT, and then Belize (Belize City is close to the center line), Honduras, and Nicaragua. It will skirt the eastern coast of Costa Rica and descend into Panama, reaching its coast at 17.56 GMT. Colombia will be the next country to see the event, the path passing south of Bogota on its way to Brazil, which it will reach at 18.56 GMT. Having crossed Brazil, the event will reach its east coast south of Natal. Very soon afterwards the eclipse will end in the Atlantic.

Total Solar Eclipse of 2024 Apr 08

Geocentric Conjunction = 18:36:02.5 UT   J.D. = 2460409.275029
Greatest Eclipse = 18:17:13.1 UT   J.D. = 2460409.261957

Eclipse Magnitude = 1.0565   Gamma = 0.3432

Saros Series = 139   Member = 30 of 71

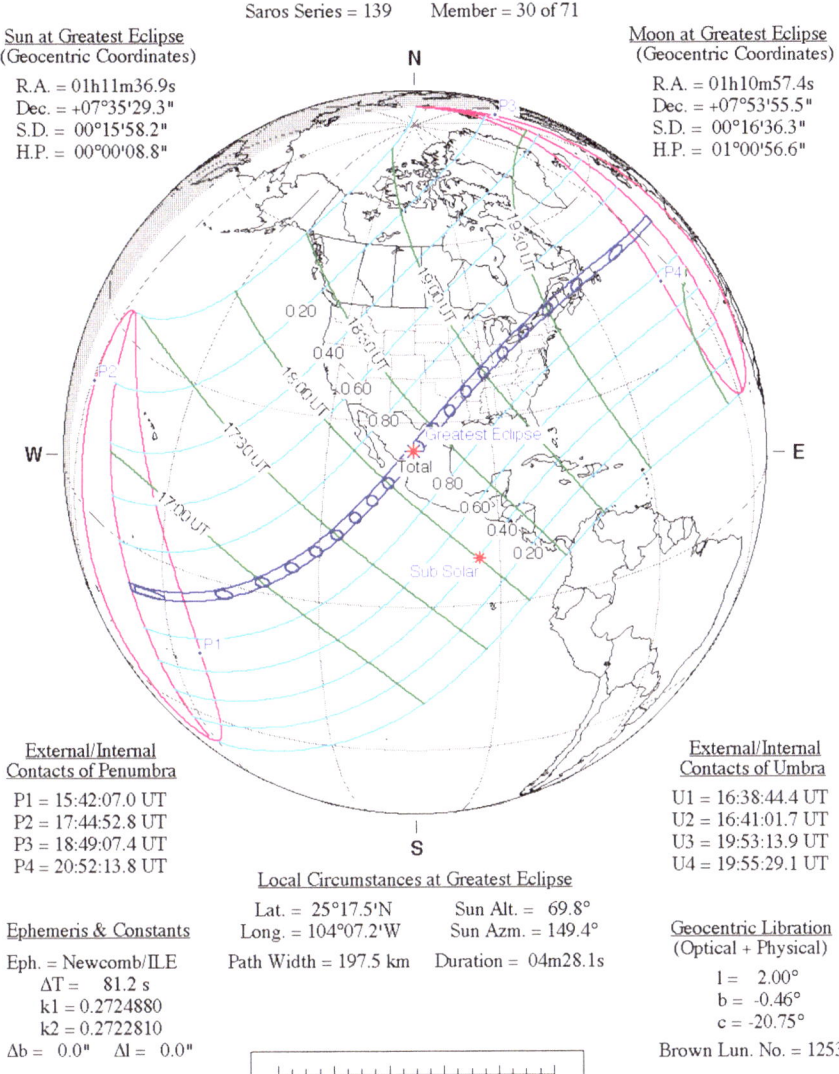

Sun at Greatest Eclipse
(Geocentric Coordinates)

R.A. = 01h11m36.9s
Dec. = +07°35'29.3"
S.D. = 00°15'58.2"
H.P. = 00°00'08.8"

Moon at Greatest Eclipse
(Geocentric Coordinates)

R.A. = 01h10m57.4s
Dec. = +07°53'55.5"
S.D. = 00°16'36.3"
H.P. = 01°00'56.6"

External/Internal
Contacts of Penumbra

P1 = 15:42:07.0 UT
P2 = 17:44:52.8 UT
P3 = 18:49:07.4 UT
P4 = 20:52:13.8 UT

Ephemeris & Constants

Eph. = Newcomb/ILE
  ΔT =   81.2 s
  k1 = 0.2724880
  k2 = 0.2722810
Δb =   0.0"   Δl =   0.0"

Local Circumstances at Greatest Eclipse

Lat. = 25°17.5'N   Sun Alt. =   69.8°
Long. = 104°07.2'W   Sun Azm. = 149.4°
Path Width = 197.5 km   Duration = 04m28.1s

External/Internal
Contacts of Umbra

U1 = 16:38:44.4 UT
U2 = 16:41:01.7 UT
U3 = 19:53:13.9 UT
U4 = 19:55:29.1 UT

Geocentric Libration
(Optical + Physical)

l =   2.00°
b = -0.46°
c = -20.75°

Brown Lun. No. = 1253

F. Espenak, NASA's GSFC - Fri, Jul 2,
sunearth.gsfc.nasa.gov/eclipse/eclipse.html

This eclipse could be one of the greats, with a shadow path up to 200 km/125 miles wide, since the Moon will be only one day past perigee. This will make its disk apparently very large compared with the Sun's. The event will be experienced across northwest Mexico, and in the United States in a diagonal line from Texas to Maine, with the last stage in eastern Canada. Maximum eclipse will occur at 18.17 GMT near Nazas, in the Mexican state of Durango. This will last for 4 minutes and 28 seconds.

A partial eclipse will occur over most of North America and in Central America. The eclipse will begin at dawn at 16.39 GMT south of Starbuck Island in the South Pacific. Isla Socorro (Mexico) will see the total eclipse, and landfall on the mainland in Mexico will occur just south of Mazatlán, where the path will attain the maximum width of 200 km at 18.09 GMT. Here, the duration of the eclipse will be 4 minutes and 27 seconds. At 18.30 GMT (13.30 CDT) the eclipse will reach the international border at Piedras Negras (Mexico) and Eagle Pass (USA). San Antonio, Austin and Dallas/Fort Worth (Texas) will be wholly or partly within the path, and also Little Rock (Arkansas), Indianapolis (Indiana), Cleveland (Ohio) and Buffalo (New York). Having crossed Vermont (with Montreal, to the north in Canada, at the northern edge of the path), the shadow will traverse Maine and leave the continent at the coast of New Brunswick south of Bathurst—just south of which, within the path, is the happily (but at that time incorrectly) named Sunny Corner. After crossing Newfoundland, the shadow will move out over the Atlantic, lifting off the water at 19.55 GMT. Nimrod Lake (Arkansas) will see both this eclipse and that of August 12, 2045, some 21 years later. Carbondale (Illinois) where the duration of totality will be 4 minutes and 9 seconds, will have seen the total eclipse of August 21, 2017, seven years before.

Annular Solar Eclipse of 2024 Oct 02

Geocentric Conjunction = 19:07:53.1 UT   J.D. = 2460586.297142
Greatest Eclipse = 18:44:51.3 UT   J.D. = 2460586.281150

Eclipse Magnitude = 0.9326   Gamma = -0.3510

Saros Series = 144   Member = 17 of 70

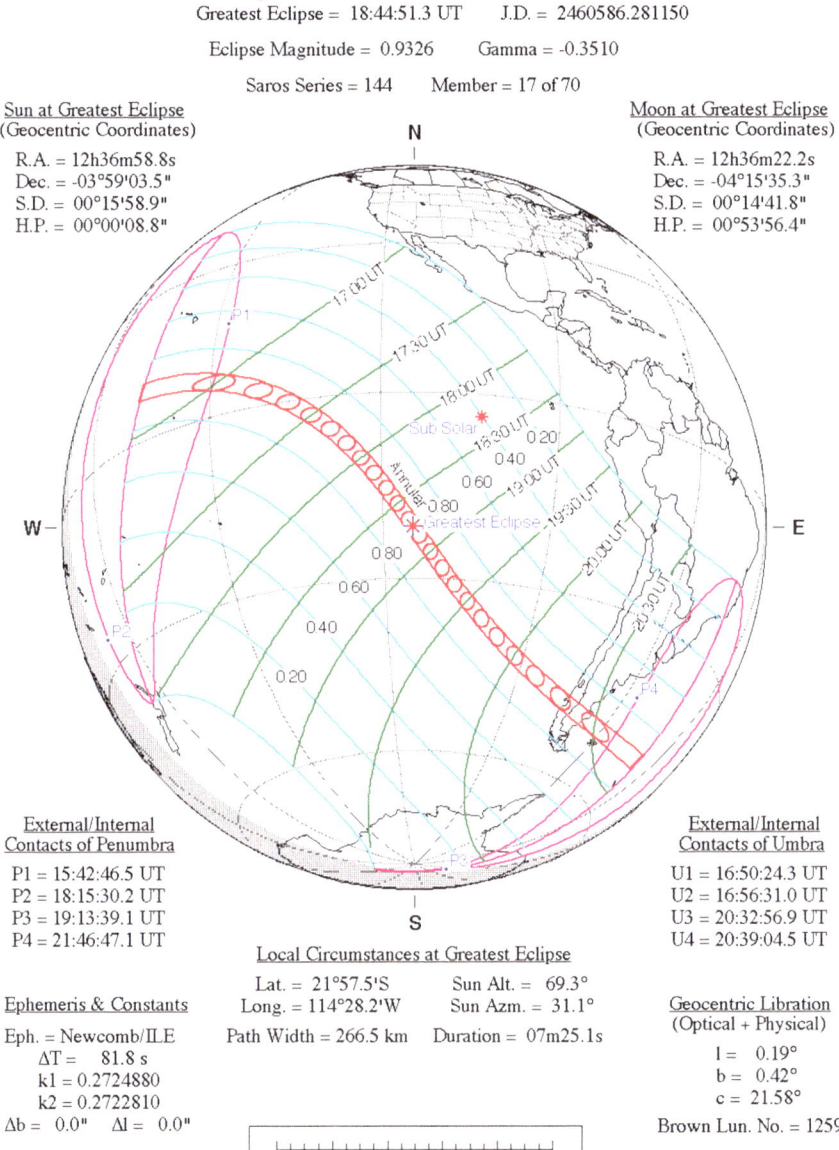

Sun at Greatest Eclipse
(Geocentric Coordinates)

R.A. = 12h36m58.8s
Dec. = -03°59'03.5"
S.D. = 00°15'58.9"
H.P. = 00°00'08.8"

Moon at Greatest Eclipse
(Geocentric Coordinates)

R.A. = 12h36m22.2s
Dec. = -04°15'35.3"
S.D. = 00°14'41.8"
H.P. = 00°53'56.4"

**External/Internal
Contacts of Penumbra**

P1 = 15:42:46.5 UT
P2 = 18:15:30.2 UT
P3 = 19:13:39.1 UT
P4 = 21:46:47.1 UT

**External/Internal
Contacts of Umbra**

U1 = 16:50:24.3 UT
U2 = 16:56:31.0 UT
U3 = 20:32:56.9 UT
U4 = 20:39:04.5 UT

Local Circumstances at Greatest Eclipse

Lat. = 21°57.5'S   Sun Alt. = 69.3°
Long. = 114°28.2'W   Sun Azm. = 31.1°
Path Width = 266.5 km   Duration = 07m25.1s

Ephemeris & Constants

Eph. = Newcomb/ILE
ΔT = 81.8 s
k1 = 0.2724880
k2 = 0.2722810
Δb = 0.0"   Δl = 0.0"

Geocentric Libration
(Optical + Physical)

l = 0.19°
b = 0.42°
c = 21.58°

Brown Lun. No. = 1259

0   1000   2000   3000   4000   5000
Kilometers

F. Espenak, NASA's GSFC - Fri, Jul 2,
sunearth.gsfc.nasa.gov/eclipse/eclipse.html

This will be a primarily South Pacific annular eclipse, touching down just north of the equator at 15.42 GMT in the center of the ocean near the International Date Line. The Moon will be at apogee at this time, making its disk apparently smaller than average and unable to cover the Sun. Only 93 % of the Sun will be hidden at maximum eclipse. The very broad shadow path, up to 266 km/165 miles wide, will cross the South Pacific, reaching Hanga Roa, on Easter Island (Chile) at about 19.05 GMT, where the annular phase will last for 7 minutes and 25 seconds, a very long period for any eclipse. Observers on the Chilean mainland will first see the eclipse at 20.23 GMT in the coastal Laguna San Rafael National Park, and it will arrive at the border with Argentina at 20.24 GMT. At 20.28 GMT the eclipse moves onto the south Atlantic, passing just north of the Falkland Islands and completing its course about 1,000 km further east.

Partial Solar Eclipse of 2025 Mar 29

Geocentric Conjunction = 11:46:09.2 UT   J.D. = 2460763.990384
Greatest Eclipse = 10:47:18.4 UT   J.D. = 2460763.949519

Eclipse Magnitude = 0.9361   Gamma = 1.0405

Saros Series = 149   Member = 21 of 71

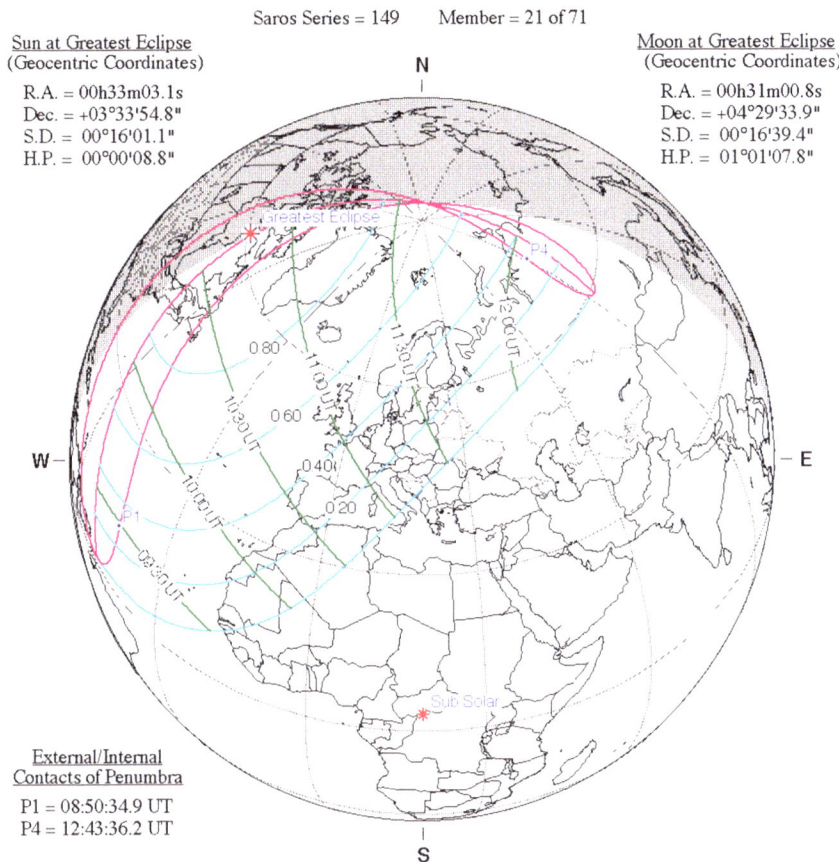

Sun at Greatest Eclipse
(Geocentric Coordinates)

R.A. = 00h33m03.1s
Dec. = +03°33'54.8"
S.D. = 00°16'01.1"
H.P. = 00°00'08.8"

Moon at Greatest Eclipse
(Geocentric Coordinates)

R.A. = 00h31m00.8s
Dec. = +04°29'33.9"
S.D. = 00°16'39.4"
H.P. = 01°01'07.8"

External/Internal
Contacts of Penumbra

P1 = 08:50:34.9 UT
P4 = 12:43:36.2 UT

Ephemeris & Constants

Eph. = Newcomb/ILE
ΔT = 82.3 s
k1 = 0.2724880
k2 = 0.2722810
Δb = 0.0"   Δl = 0.0"

Geocentric Libration
(Optical + Physical)

l = -2.00°
b = -1.35°
c = -21.73°

Brown Lun. No. = 1265

0   1000 2000 3000 4000 5000
Kilometers

F. Espenak, NASA's GSFC - Fri, Jul 2,
sunearth.gsfc.nasa.gov/eclipse/eclipse.html

If the center of the Moon's shadow does not touch Earth, as can occur in regions near the poles where the cone-shaped lunar umbral shadow grazes the planet, an eclipse will be partial across all its range. The partial eclipse of the Sun of March 29, 2025, will be experienced from the northeastern corner of North America, Greenland, Iceland, western and northern Europe and northwest Africa. The spectacle will begin at 08.50 GMT and last until 12.43 GMT. This partial eclipse will be a substantial one, and luckier viewers will see, with appropriate eye protection, up to 94 % of the Sun covered in the far north. This will also be a significant spectacle for viewers in the UK, with about half of the Sun hidden.

Partial Solar Eclipse of 2025 Sep 21

Geocentric Conjunction = 20:50:18.4 UT   J.D. = 2460940.368269
Greatest Eclipse = 19:41:43.6 UT   J.D. = 2460940.320643

Eclipse Magnitude = 0.8535   Gamma = -1.0652

Saros Series = 154   Member = 7 of 71

Sun at Greatest Eclipse
(Geocentric Coordinates)

R.A. = 11h56m36.8s
Dec. = +00°22'01.0"
S.D. = 00°15'55.9"
H.P. = 00°00'08.8"

Moon at Greatest Eclipse
(Geocentric Coordinates)

R.A. = 11h54m42.7s
Dec. = -00°29'14.7"
S.D. = 00°15'02.8"
H.P. = 00°55'13.2"

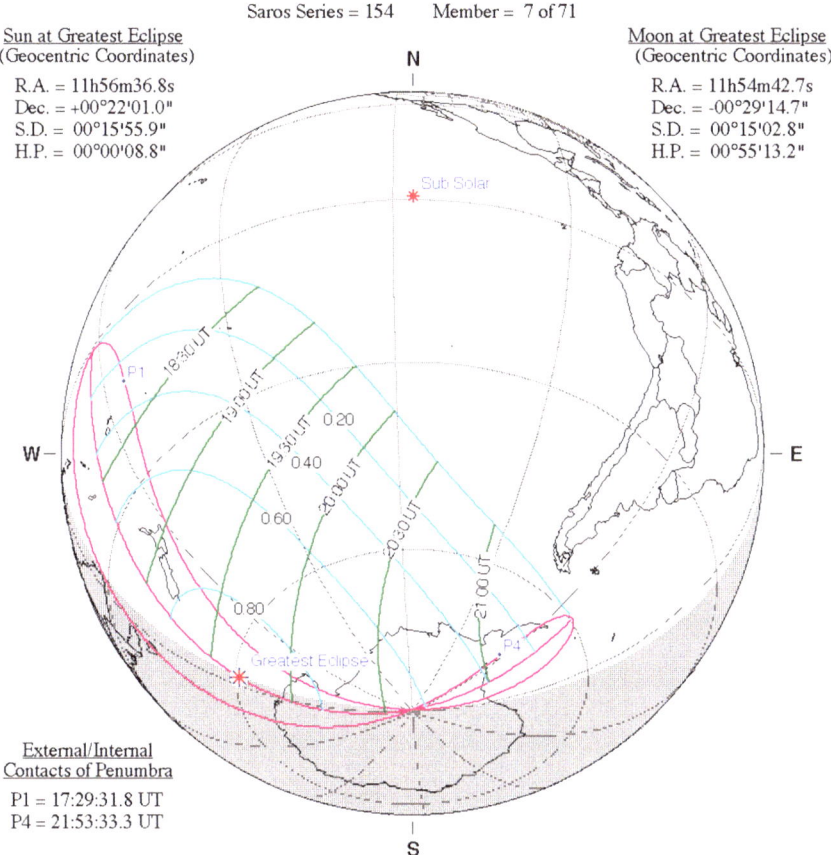

External/Internal
Contacts of Penumbra

P1 = 17:29:31.8 UT
P4 = 21:53:33.3 UT

Ephemeris & Constants

Eph. = Newcomb/ILE
ΔT = 82.8 s
k1 = 0.2724880
k2 = 0.2722810
Δb = 0.0"   Δl = 0.0"

Geocentric Libration
(Optical + Physical)

l = 4.15°
b = 1.31°
c = 21.92°

Brown Lun. No. = 1271

0   1000   2000 3000 4000 5000
Kilometers

F. Espenak, NASA's GSFC - Fri, Jul 2,
sunearth.gsfc.nasa.gov/eclipse/eclipse.html

This second partial solar eclipse of 2025 occurs in the far south. It will be substantial (with up to 80 % of the Sun covered) in New Zealand, and will move down onto western Antarctica and the Antarctic peninsula.

Annular Solar Eclipse of 2026 Feb 17

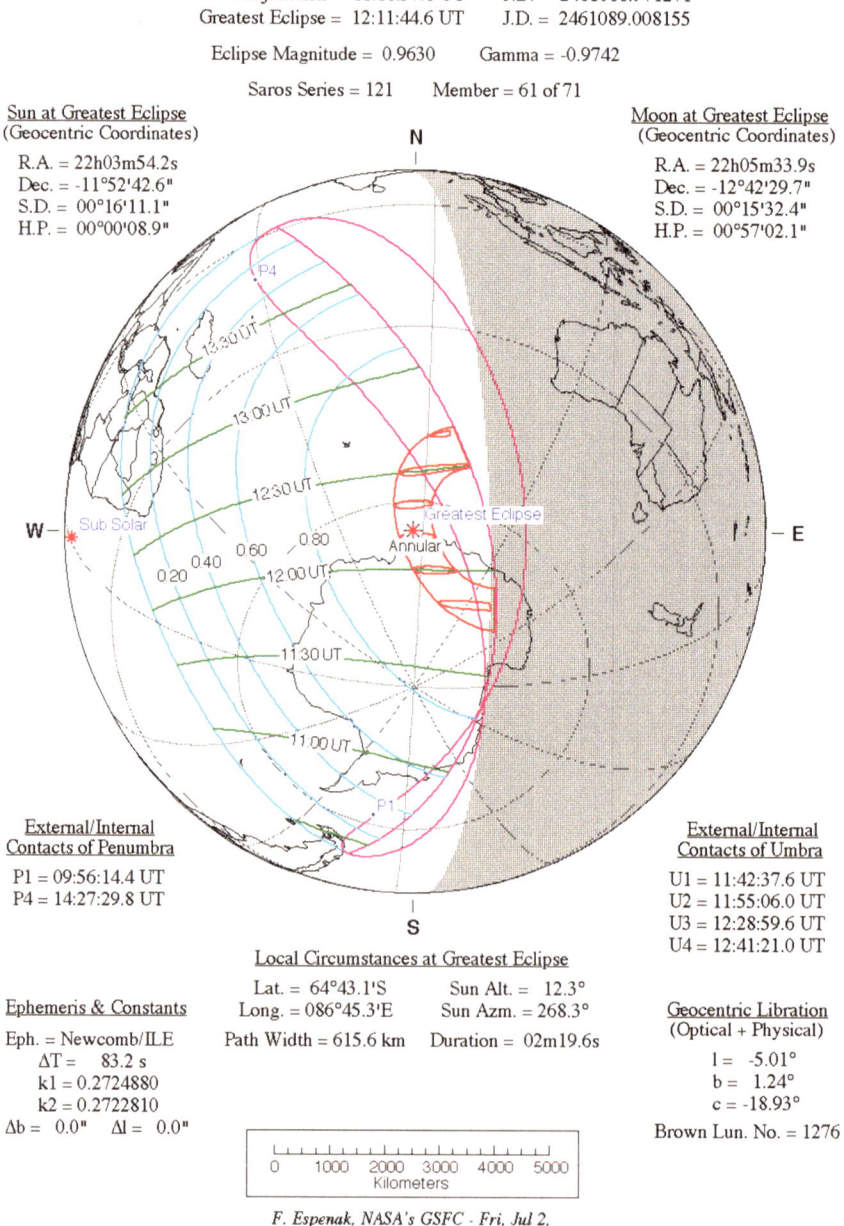

Geocentric Conjunction = 11:18:37.8 UT   J.D. = 2461088.971271
Greatest Eclipse = 12:11:44.6 UT   J.D. = 2461089.008155

Eclipse Magnitude = 0.9630   Gamma = -0.9742

Saros Series = 121   Member = 61 of 71

Sun at Greatest Eclipse
(Geocentric Coordinates)

R.A. = 22h03m54.2s
Dec. = -11°52'42.6"
S.D. = 00°16'11.1"
H.P. = 00°00'08.9"

Moon at Greatest Eclipse
(Geocentric Coordinates)

R.A. = 22h05m33.9s
Dec. = -12°42'29.7"
S.D. = 00°15'32.4"
H.P. = 00°57'02.1"

External/Internal Contacts of Penumbra

P1 = 09:56:14.4 UT
P4 = 14:27:29.8 UT

External/Internal Contacts of Umbra

U1 = 11:42:37.6 UT
U2 = 11:55:06.0 UT
U3 = 12:28:59.6 UT
U4 = 12:41:21.0 UT

Local Circumstances at Greatest Eclipse

Lat. = 64°43.1'S   Sun Alt. =   12.3°
Long. = 086°45.3'E   Sun Azm. = 268.3°
Path Width = 615.6 km   Duration = 02m19.6s

Ephemeris & Constants

Eph. = Newcomb/ILE
ΔT =   83.2 s
k1 = 0.2724880
k2 = 0.2722810
Δb = 0.0"   Δl = 0.0"

Geocentric Libration
(Optical + Physical)

l = -5.01°
b =   1.24°
c = -18.93°

Brown Lun. No. = 1276

F. Espenak, NASA's GSFC - Fri, Jul 2,

sunearth.gsfc.nasa.gov/eclipse/eclipse.html

The annular solar eclipse of February 17, 2026, will be visible (if there's anyone hardy enough to go there) from a wide swathe of eastern Antarctica, the maximum coverage of the Sun's disk occurring just above the coastline. The path will then curve across part of the Atlantic Ocean in the direction of Australia, lifting off about 2,000 km/1,250 miles from the Antarctic. The associated partial eclipse will be visible from the very southern tip of South America, most of Antarctica, much of the Atlantic Ocean, parts of southwestern Africa, Madagascar and the southwestern Indian Ocean.

Total Solar Eclipse of 2026 Aug 12

Geocentric Conjunction = 17:03:39.9 UT   J.D. = 2461265.210878
Greatest Eclipse = 17:45:43.7 UT   J.D. = 2461265.240089

Eclipse Magnitude = 1.0386   Gamma = 0.8976

Saros Series = 126   Member = 48 of 72

Sun at Greatest Eclipse
(Geocentric Coordinates)

R.A. = 09h29m47.2s
Dec. = +14°48'04.7"
S.D. = 00°15'47.0"
H.P. = 00°00'08.7"

Moon at Greatest Eclipse
(Geocentric Coordinates)

R.A. = 09h31m17.3s
Dec. = +15°36'58.2"
S.D. = 00°16'16.9"
H.P. = 00°59'45.1"

External/Internal
Contacts of Penumbra

P1 = 15:34:01.1 UT
P4 = 19:57:47.6 UT

External/Internal
Contacts of Umbra

U1 = 16:57:54.1 UT
U2 = 17:01:54.9 UT
U3 = 18:30:01.4 UT
U4 = 18:33:57.4 UT

Local Circumstances at Greatest Eclipse

Ephemeris & Constants

Eph. = Newcomb/ILE
ΔT =   83.8 s
k1 = 0.2724880
k2 = 0.2722810
Δb = 0.0" Δl = 0.0"

Lat. = 65°13.0'N   Sun Alt. =   25.8°
Long. = 025°13.6'W   Sun Azm. = 248.3°
Path Width = 293.8 km   Duration = 02m18.3s

Geocentric Libration
(Optical + Physical)

l =   4.08°
b = -1.12°
c = 16.98°

Brown Lun. No. = 1282

F. Espenak, NASA's GSFC - Fri, Jul 2,
sunearth.gsfc.nasa.gov/eclipse/eclipse.html

This will be the first European total solar eclipse since 2015, and this time across part of the mainland instead of Atlantic and Arctic islands. Totality will be seen from part of eastern Greenland, the west of Iceland (including Reykjavik, where totality will last for only a few seconds because it will be very close to the eastern edge of the shadow track), the western Atlantic Ocean, Spain, a very small area of northern Portugal at the edge of the track, and the Balearic Islands in the western Mediterranean. The partial phase will be seen from Canada and parts of the northern United States, western Europe and North Africa. The lunar shadow touches down in the Arctic on the northern coast of Siberia and moves onto the Arctic Ocean, curving towards Greenland. The greatest duration of this eclipse will occur as seen from the sea near the west coast of Iceland. At maximum the Sun will be blotted out for 2 minutes and 18 seconds, and its very wide path will be at most nearly 300 km/186 miles across. This will be the first total eclipse seen from Iceland since June 30, 1954, but Icelanders will have to venture out to sea from the coast to a point 230 km/143 miles northwest of Reykjavik to get the best of it. Three-quarters of an hour later the shadow will pass over the west coast of northern Spain, making landfall near Oviedo at about 18.28 GMT (19.28 Spanish time).

This will be the first total eclipse in Spain since 1905, when the path was similar. It will cross from the west coast to the east, then move across the Balearic Islands of Minorca, Majorca and Ibiza. Weather permitting, totality will be visible from several Spanish cities, including, in chronological order, León, Santander, Valladolid, Bilbao, Burgos, Zaragoza and Valencia. Both Madrid and Barcelona will lie just outside the shadow's path, as will Pamplona and Santiago de Compostela. A very small part of northern Portugal, within the Montesinho Natural Park, will be briefly within the lunar shadow, the maximum duration of the event in this area being up to 24 seconds. From Ireland, the UK, the rest of Spain and Portugal, France, Italy, the Balkans and much of northwest Africa the partial eclipse will cover more than 90% of the Sun, with lesser coverage for elsewhere in Europe, further east in North Africa, and North America. Interestingly, Spain will not have to wait long after this for its next total eclipse, which will visit the country less than a year later on August 2 2027.

Geocentric Conjunction

The Moon is said to be in geocentric conjunction with the Sun at the point in the lunar cycle commonly known as the new Moon: the moment at which the centers of the Earth, Moon and Sun are in line with the Moon between Earth and the Sun. The time given for this instant in the notes is expressed, as is normal in astronomical literature, in Universal Time (UT), for everyday purposes the same as Greenwich Mean Time (GMT). For conversion of UT to the time in any time zone, see earthsky.org/astronomy-essentials/universal-time.

J. D. (Julian Day)

I'm writing this on Julian Day number 2457473, otherwise known as March 26, 2016. In normal circumstances it isn't at all necessary to know what Julian Day it is, but anyone interested in the curious and complex history of calendars and humanity's struggle to sort it all out might like to look this up. Try scienceworld. wolfram.com/astronomy/JulianDate.html.

Greatest Eclipse

This is defined as the moment when the central axis of the Moon's cone-shaped shadow (assumed for the purposes of computation to be circular) passes closest to the center of Earth. Because of the presence of mountains and valleys at the lunar limb ('limb' being the astronomical term for the edge of a celestial body), the disk of the Moon is not a circle. The Moon's shadow is about 380,000 km/ 236,000 miles in length, around 108 times as long as its diameter. This is normally long enough for the shadow to reach Earth, causing a total eclipse. If the Moon is further from Earth than usual and the shadow cannot reach it, there will be an annular eclipse.

Eclipse Magnitude

This refers to the fraction of the diameter of the eclipsed Sun or Moon which is actually in eclipse, in the case of the Sun by the Moon's disk, and in the case of the Moon by Earth's shadow. So a partial eclipse, or a 'ring-of-fire' annular eclipse, will have a magnitude of less than 1.0, for example in the case of the annular eclipse shown in the diagrams above for February 17, 2026, where the figure is 0.963.

Saros Series

The ancient Chaldeans noticed, after hundreds of years' observing the cycles of celestial bodies (thought to be deities in the sky) that series of lunar eclipses seemed to repeat themselves. This periodicity is governed by the Saros cycle, a term said to have been coined by the great English mathematician, astronomer and meteorologist Edmond Halley. It is a period of approximately 6,585.3 days, or 18 years 11 days and 8 hours. It can be applied to solar eclipses as well. This enables us to predict patterns of eclipse paths based on those which have happened in the past. For further explanation of this complicated subject see https://en.wikipedia.org/wiki/Saros (astronomy).

Contacts

During an eclipse, astronomers describe the progress of the event in terms of contacts. The first contact (U1) occurs at the beginning of the partial phase of the eclipse, when, in the case of a solar eclipse, the Moon's disk starts to encroach upon the Sun's. Second contact (U2) is at the moment when the Sun is completely eclipsed and totality begins. Third contact (U3) occurs when totality ends and the Sun reappears. Fourth contact (U4) occurs when the Moon leaves the solar disk completely. 'U' refers to the fact that it is the umbral shadow of the Moon that is in question, not its larger penumbral shadow. These terms are also used to record the progress of Earth's shadow across the Moon during a lunar eclipse.

Sub Solar (Subsolar) Point

This refers to the point on Earth (or indeed any other planet if there were any observer there) where the Sun is at the zenith, the position directly overhead. During the day the subsolar point moves westward, and in the course of a year it will also gradually move north and south between the Tropic of Cancer (23°26′ North) and the Tropic of Capricorn (23°26′ South). At the equinoxes in March and September the subsolar point crosses the equator.

Chapter 6

The Star-Hunter's Kit

So what is the best telescope for you? The simple answer is: it's the one that you think you'll use the most often.

—Steve Tidey, Association for Astronomy Education

I have looked farther into space than ever a human being did before me.

—William Herschel

It's not one of the aims of this book to compare telescopes and binoculars and recommend best buys. Visiting retailers and talking to local astronomers are the best ways forward, but let's discuss in general terms what the beginner should be looking for if planning to travel somewhere to observe with some optical aid. You'll probably come to the conclusion that what matter are three Ps: portability, price, performance.

Many people have said something like: "I bought a telescope a couple of years ago, quite an expensive one in fact, but I just couldn't get the hang of operating it. It's in the garage now, gathering dust. I want to sell it." A sadder story is that they have bought a child a telescope hoping to ignite some interest in science, and "unfortunately (s)he couldn't find anything but the Moon."

First, astronomy without a telescope or binoculars can be a really enjoyable experience. You have more time to observe if you're not unpacking, assembling, adjusting or maneuvering instruments. Relaxing beneath the stars with an unrestricted field of view, your unaided eyes may notice meteors, satellites, the tip of the Moon rising, a possible auroral glow starting up in some far corner of the sky—it can be a busy

© Springer International Publishing Switzerland 2016
B. Mizon, *Finding a Million-Star Hotel*, The Patrick Moore Practical Astronomy Series,
DOI 10.1007/978-3-319-33855-2_6

place. You won't have the frustrations of keeping dew off instruments, searching for invisible lens caps in the dark (like Model T Fords, they always seem to be black!) or having to pack up your kit at the end of a session when you'd rather be on the way home to bed.

Learn the sky. Before you even contemplate buying a telescope or binoculars, learn the night sky from charts, night-sky simulators (see Appendix 4 in this book) and through real outdoor observation. All those thousands of dots seem a bit daunting, but, like learning a foreign language, the best way in is immersion. Using a plani-sphere (rotating sky chart), a laminated star map or an electronic guide on a phone or other portable device, get out under the stars as often as you can. It won't take long to master the basic constellations, work out how the Moon and planets move, and learn the locations of a few deep-sky objects such as the Andromeda Galaxy.

Next, get binoculars. You will now be filling in the gaps in your knowledge of the night sky, finding more objects and preparing the way for eventually zeroing in on them with a telescope. This doesn't mean a telescope will necessarily take the place of binoculars—they both have their advantages and can be used together. Remember if you decide to do this that while astronomical telescopes turn things upside down, binoculars don't.

Common sizes for lightweight binoculars are 7×50, 8×40 and 10×50. These numbers simply mean that the magnifications will be 7, 8 or 10 times, and the aperture or distance across the objective (larger) lenses is 40 or 50 mm. The larger the aperture, the greater the amount of light entering the binoculars, and the fainter the objects seen.

Of course, higher magnification lets you see more detail, and also darkens the sky background, improving contrast. This is important for extended faint objects. It's not just a case of the bigger the better, however. Increasing aperture means you are increasing the size and the weight of your instrument, and hand-holding becomes increasingly problematical. Even lightweight instruments cannot be held comfortably for long. You may have steady hands, but the muscles in your arms will soon start protesting, and an uncomfortable observer is not a successful observer. A tripod or monopod (Fig. 6.1) is a good solution to this problem.

Probably the best size of binocular for the beginner is the 10×50. These are easily carried around, and if dangling on a neck strap or carried in their case they will not be too burdensome. They fit into car glove compartments, handbags and large pockets. They are not too heavy in use, even for small children, though supporting them is not difficult. Even a stout pole driven into the ground can be an acceptable mount if nothing else is available. They have enough aperture (light-gathering capability) to show to good advantage beautiful open clusters such as the Pleiades, or fainter objects such as the Andromeda Galaxy. They can be used for daytime activities such as bird watching. Don't be tempted by bargains. Some large discount supermarkets now offer binoculars at very low prices, especially before Christmas. If they cost less than the rest of your shopping, be suspicious! You get what you pay for, and it's better to spend a bit more, after visiting retailers and taking advice from local astronomers or websites such as www.binocularsky.com (UK) and www.skyandte-lescope.com/astronomy-equipment/binoculars-for-astronomy (United States).

Fig. 6.1 Binoculars supported on a monopod. (Photo: Steve Tonkin)

The 10×50's smaller cousin, the 7×50, shows some increase in the brightness of extended objects such as nebulae, but they also increase background sky brightness, so in practice they give negligible advantage except for only a few larger faint objects. The main plus-point of the 7×50 is that it is easier to hold steadily and for longer periods.

Larger instruments such as 15×70s and 20×80s always need supporting. They are too heavy to hold comfortably. Many binoculars have a screw-socket and can be adapted, with a suitable bracket, to fit on a standard photographic tripod. There are alternative mounts. Increasingly popular among astronomers is the parallelogram mount (Fig. 6.2), which enables observers to look through the binoculars without the use of hands, and also adjust the height of the instrument for different people (invaluable at star parties with a mixture of children and adults) without losing the object they are pointing at. It keeps it in the field as the binoculars are moved up and down.

When you feel you know the sky well enough, and can use a detailed star chart confidently, you might take the plunge and buy that telescope. As Steve Tonkin, one of the UK's most respected authorities on astronomical instruments, says: "You probably don't have a lot of money to spend, and you want to get the best telescope

Fig. 6.2 Binoculars on a parallelogram mount. (Photo: Steve Tonkin)

for your money. You also want to get a telescope that lets you get the most out of the night sky—one that lets you see everything you want to see."

The hard reality is that there is no such thing as an all-around, see-everything telescope. The most important thing is to find one that is easy to use, so for the beginner the "all-singing, all-dancing" motorized kind with computer handset, bulky tripod and lots of spiraling wires might not be best. A medium-sized refractor, the "straight-through" kind of telescope with a lens at either end, is usually fairly light and comparatively easy to aim. Looking at stars high above through a refractor can be problematical. You have to get almost underneath the telescope to look through the eyepiece. This is solved by using a right-angled eyepiece holder called a diagonal. A refractor will probably have a small finderscope or red-dot finder attached to it to help you line it up on the object you want to look at.

A word of warning—if the telescope box in the store shows pictures in glorious color of nebulae and galaxies, don't be deceived. You will not see such colors through your telescope, and those photos are false advertising. The Orion Nebula will, for example, look faintly greenish-gray, and the Ring Nebula (Fig. 6.3) will resemble a gray smoke-ring. The colors in the photos merely underline the fact that a camera can store and reproduce information, and your eyes cannot.

What the telescope sits on can be as important as the quality of its optics. A firm tripod or other mount is absolutely essential. Stars, planets, the Moon and faint

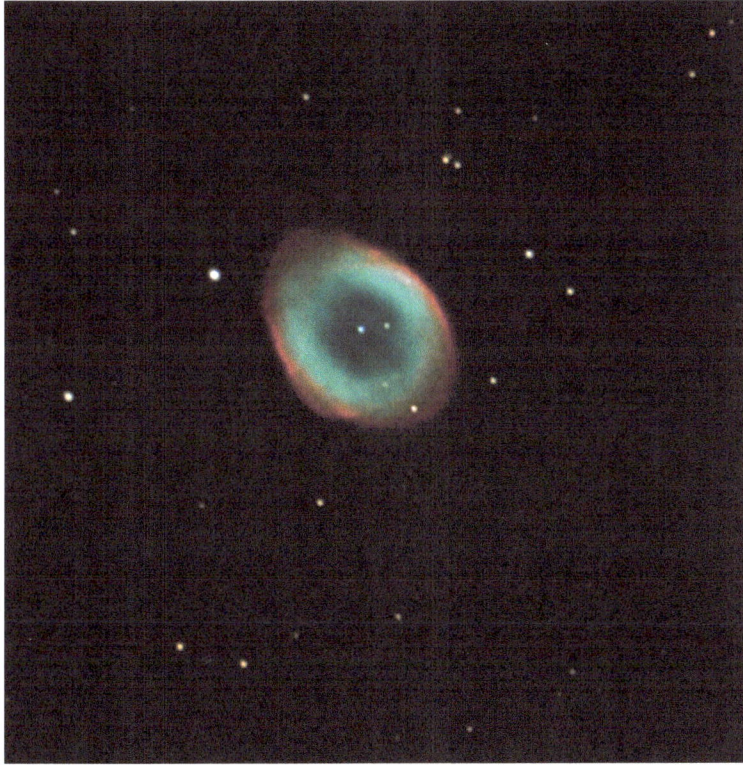

Fig. 6.3 M57, the Ring Nebula in Lyra. (Photo: Neill Mitchell)

galaxies are no fun at all if they are vibrating in the breeze or veering around in the telescope field.

So a simple, easy to use set-up is the thing to aim for, but don't be fooled into assuming that simple means cheap. Be prepared to invest in good quality—and seek advice.

Now, what about reflectors, the kind of telescopes that use a mirror to catch light and reflect it back to the eyepiece? Is there a reflector that is portable, easy to operate and lets you set it up and find things quickly? Yes. It's the Dobsonian telescope, named after its genial inventor John Dobson (1915-2014), co-founder of the San Francisco Sidewalk Astronomers. A telescope tube that pivots up and down, sitting on a rotating box (Fig. 6.4)—as simple as that, able to be pointed anywhere in the sky without electronics. You just haul it around and aim using a finderscope or red-dot device. You can take it to pieces for easy transportation. By all means graduate to something a bit more 'techno' later, but you might get hooked on Dobs. Many astronomers have used mostly Dobsonians and enjoy the easy success in finding things in the sky—the thrill of the star-chart chase followed by the glow of satisfaction

Fig. 6.4 A trolley-mounted 14-inch Dobsonian telescope. (Photo: Bob Mizon)

from focusing in on, for example, a spindly edge-on galaxy 30 million light years away with a faintly discernible dust lane cutting across its central bulge (Fig. 6.5).

A brief internet search will turn up companies who hire out instruments if you prefer not to buy.

There are many good star atlases on the market, some suitable for beginners with basic stars and telescopic objects, and others with tens of thousands of stars, nebulae and galaxies for the experienced telescope user. Star charts, often laminated against the damp, are available from astronomical suppliers and organizations, booksellers and Internet retailers. Make sure they have clear print that's easy to see with a dim red light.

Planispheres have rotating discs that allow you to set the date against the time and display the current night sky. There are websites giving instructions on making your own.

Fig. 6.5 The "Spindle Galaxy," NGC 891, an edge-on spiral in Andromeda. (Photo: Ian Smith)

It is surprising how easy it is to keep cold air out. Wear several loose layers of clothing to trap warmth around your body. The neck and head and the small of the back should be well protected. Stand on a mat. Invest in small warming devices for the hands and pocket.

Don't drink too much fluid, for obvious reasons. Stick to a little water if thirsty. Alcohol doesn't warm—it dilates the blood vessels under the skin, drawing blood away from inner vital organs towards the exterior, where heat can be lost more quickly.

Other items observers can keep in their bags are a red lamp, though you can get pens with built-in red lights, insect repellent (don't spray near optical instruments!), a folding stool, a kneeling pad, and a modest white light for clearing up after observing or for finding dropped items.

Smartphones and other electronic devices have become useful allies in the star-seeker's arsenal. Locate and identify constellations, bright stars, planets and satellites with a whole range of apps and programs too numerous to mention here. Listings of such helpful guides are on websites such as www.iphoneness.com/iphone-apps/top-astronomy-applications-for-iphone or www.skyandtelescope.com/astronomy-resources/astronomy-software-public-domain-freeware-and-shareware.

Many of these have neat features, such as displaying southern hemisphere stars if you point them at the ground or zeroing in on objects to display them as if seen through a telescope.

Chapter 7

Stargazing Etiquette and Hosting Astro-Tourism

Star-gazing is a quiet, peaceful activity.

–Go-astronomy.com

What's the right etiquette for group stargazing? How can you make the experience as pleasurable as possible for yourself and others? The following section provides some useful tips.

- Most large modern telescopes are mounted on tripods that often extend quite a way beyond the instrument itself. Owners may also have objects on the ground such as power packs or batteries to drive their telescopes. Move slowly in the vicinity of instruments and watch your step. If you're a telescope operator, make sure everything is safely stowed or kept beneath the tripod rather than where other observers walk.
- Be especially careful of wires trailing on the ground. There may also be people on the ground! Some observers simply want to relax and enjoy the night sky above, and the simplest way is just to lie down on a blanket and look up. If this is your preferred method, stay on the periphery of the gathering!
- Telescope users should provide some kind of secure platform for children to stand on so they can comfortably get their eyes to the eyepiece.

© Springer International Publishing Switzerland 2016
B. Mizon, *Finding a Million-Star Hotel*, The Patrick Moore Practical Astronomy Series,
DOI 10.1007/978-3-319-33855-2_7

- Optical instruments are delicate and expensive. Try not to touch them or lean on them while looking though eyepieces. It will have taken time to correctly align motor-driven telescopes, and owners want them to stay put.
- If binoculars are on loan, keep them with you. Use the neck-strap if they have one. For whatever reason, they are usually black, so leaving them on the ground in the dark is not a good idea.

- It's not easy to concentrate on a tiny point of light, for example one of Saturn's moons seen through a 10-inch telescope, if someone near you is listening to a rock concert through leaky earphones. Give your portable music a rest for a night. If someone near you points out an interesting happening in the sky, for example the International Space Station passing over, you'll probably want to hear the announcement rather than your music.
- Keep conversations low and unobtrusive. This is not only common courtesy, but is also advisable since someone might want to announce that a satellite is crossing the sky, or that a lost child is looking for his or her parents.

- Don't use bright lights while observing is going on. Red lights are essential, the dimmer the better, within reason. If brighter light really is needed, for example to help that lost child or locate equipment, announce the fact to those near you that you are going to switch it on.
- It's probably better not to bring laptops and tablets to the observing site, not just because the light from them can degrade night vision but because someone (even the owner) might just blunder into them or tread on them in the dark when they're not displaying. The damp conditions often prevailing at night might not be good for them either. Phones, too, can spoil your night vision, and that of people nearby. Many night sky programs and phone apps have night modes or can be filtered with suitable red translucent material, so you can use them yet keep your eyes ready to observe.
- Vehicle lights can create problems if cars move near the observing area. You could park at a distance and walk in if you're not heavily laden, or even better, get there before dark. If you approach observers at night, drive *very* slowly using sidelights only. Organizers should appoint a steward to ensure safe vehicle movement. It's a good idea to think vehicle lighting through in daylight before you set off. Look in the vehicle's manual for information if you don't already know how to disable the interior lights when a door is opened.

- There's a current debate about the use of laser 'pens' or torches/flashlights as sky pointers. Many star gatherings ban them. Flashlights are always inappropriate as pointers and should never be used for the purpose. Laser pointers have concentrated beams and can be very useful and accurate educational tools, in a steady hand, for identifying individual planets, stars and constellations, though the Moon is pretty self-evident! Interestingly, the highly directional beam of a laser pointer scatters preferentially forwards and backwards, not sideways, from the dust and water droplets in the air. Consequently those looking at it from beneath see the beam easily, but from a certain distance, it will be invisible. To use or not to use? Here's a personal opinion. Laser pointers ought to be used sparingly. One per star party, handled only by organizers who know not to point it near people, animals, aircraft, or telescopes—this is illegal everywhere, and also dangerous!

  A good idea is to hold a couple of "sky tour" sessions early on during the proceedings, with observers grouped around the organizer holding the pointer—if they're not near the user the beam will appear less bright. He/she should explain to the audience that they will keep the beam well away from any living thing. People sometimes do point them at trees, for example, to demonstrate the concentrated nature of the beam, forgetting that trees are homes to many a living creature. So, restricting the use of pointers to trained staff members and expert astronomers is the best option, though banning them altogether runs a close second.
- If astrophotographers are at work, even if not on the observing site but nearby, don't allow pointers and keep all lights to a minimum. Today's highly sensitive cameras and the long exposures astrophotographers use will capture the stray light, even if the eye can't detect it from a distance.

- It should be ensured that the owner of the site, if applicable, approves your use of the property, and that any individual, authority, organization or group hosting astronomical activity at such a site has the correct arrangements in place for inspection, public liability and the protection of minors and vulnerable adults.
- A further note on laser pointers. They must conform to the laws current at the location used. A useful website on this subject for U. S. observers is www.laser-pointersafety.com/rules-general/uslaws/uslaws.html, and for the UK, www.ask-thepolice.uk/Content/Q757.htm.

Believe it or not, there are some people who are not anxious to look at the night sky. Try not to disturb non-participants, for example at a campsite where non-astronomers think a good night's sleep more important than a good night's cruising through the universe.

Lists of upcoming star parties in the United States can be found at www.skyan-dtelescope.com/astronomy-events.

Dave Eagle's informative website lists the principal star parties throughout the UK: www.eagleseye.me.uk/Sky/Wordpress/?p=3312.

> We have been working very hard with our residents and tourism businesses. Almost everyone I have dealt with has seen an increase in visitors seeking dark sky experiences, and some businesses have developed facilities for their guests to enjoy the night sky – with observatories and telescopes for customers to use.
>
> –Ruth Coulthard, Brecon Beacons National Park IDSR, Wales

Planning, Telescope Hire, Lights, Accommodation

If you're inviting people to your permanent dark sky site for the specific purpose of seeing the stars, or arranging a one-off event for would-be sky watchers, you need a plan. Take care to make the site astronomer-friendly, provide observing equipment and involve local experts to navigate the skies. Leaders and telescope owners at some events wear high visibility jackets. This can be more useful than expected. The increased visibility of personnel means that far more questions get asked and people find out a lot more about the night sky.

Ensure that everyone knows how to make the experience memorable, rather than frustrating, for themselves and others. At well-organized events, as darkness falls, everyone knows exactly what's going to happen, and those in charge really know what they're doing. Happily, this is the usual state of affairs. Most astronomers are very sane, tolerant people. Maybe this has something to do with them being astronomers.

However, it is not uncommon to go to events where lack of forethought and proper supervision has caused meltdown—fifty people expected to share two telescopes; lost children wailing; bright lights waving; laser pointers endangering eyes and illegally targeting aircraft; disputes between astronomers and sleepless non-astronomers, resulting in police intervention (yes, it really happened); and cars periodically spoiling everyone's night vision as headlights sweep across the area.

You need to inform potential observers, both in your publicity and on site, about what's available. Visitors will want to know what astronomy experiences await them, and what facilities, both astronomical and every day, can be found. Is any local exterior lighting, either on site or away from it, going to color the sky? The Commission for Dark Skies (CfDS) and the International Dark-Sky Association (IDA) offer advice on ensuring good lighting practice at:

www.britastro.org/dark-skies/guidelines.html
darksky.org/lighting/lighting-basics.

Are there plenty of things to do on cloudy nights? What about accommodation? If yours is limited, what nearby alternatives are there?

Dark Site Guidelines

The CfDS has compiled advice for dark sky locations it has worked with on organizing observing and making visitors' experiences positive, rewarding and safe. Here is a collation of that advice.

Are you promoting dark sky zones, IDA-designated or otherwise, with associated camping and caravan sites, hotels and guest houses? What should a worthwhile night-sky observing site offer?

There should be:

- Knowledgeable observers, introduced by name beforehand, to lead the proceedings. Local astronomical societies are usually happy to help. Ensure enough tele-scopes and/or mounted binoculars (preferable to hand-held instruments) for everyone to observe frequently. Space out telescopes to allow free movement. Observing with the unaided eye should also be encouraged, especially in the case of meteor watches, and this too should be expert-led.
- Dark night skies in all directions, with any visible light pollution well down towards the horizon. If fainter stars or the Milky Way can't be seen from the site, it's not a good one. A multi-directional SQM measurement of 19.50 or more is desirable. Sky measurements can often be arranged with a local astronomical group or (in the UK) the Commission for Dark Skies.
- No local/on-site lighting likely to interfere with night-sky viewing, either by shining towards observers or by creating skyglow. Nearby lights should be cor-rectly shielded and angled, or if possible turned off. Mark paths to observing sites with faint glow-sticks or similar.
- A relatively flat area on firm ground or stands, freely open to visitors, large enough for groups of observers and telescope users.
- A safe, dark environment, without ground-level obstacles, thorn bushes, pits, animal burrows or other potential foot traps.
- Minimal obstructions to viewing, such as hills, nearby tall trees and buildings.
- Good late-night access, with nearby vehicle parking not so close to the site as to allow dazzling observers with headlights. Use signs on approach roads recommend-ing sidelights and very low speeds on site.
- Facilities for wheelchair users.
- Availability of planispheres, sky charts, red lights.
- Information on site about local accommodation and retail outlets.

Basic rules for observers, condensed from those given earlier, are:

- Muted red lights only on the observing site.
- Trained staff only to use laser 'pens' as sky pointers, and always with the consent of all present. Check local laws.

- No food, drink or litter in the observing area. Ensure non-alcoholic drinks are available somewhere nearby.
- Ensure public liability insurance is in place, and plan for the protection of minors and vulnerable adults.
- The CfDS website has information on measures to reduce light pollution. The last chapter is given over to this subject, since it is, for the most part, light pollution that causes people to have to travel to see starry night skies. Light pollution is a problem with straightforward solutions that will give us all better lit towns, enhance rural tranquility, and save a very large amount of money. When travelers return from standing (or lying!) beneath the stars, they may want the optimum night sky in their home areas, and ask local decision makers to choose star-quality lighting.

Chapter 8

Light Pollution:
Thief of the Stars
and Mother of Dark
Sky Tourism

There are two kinds of light – the glow that illumines, and the glare that obscures.

–James Thurber, *Lanterns and Lances*, Harper, 1961

The Moon and the stars no longer come to the farm. The farmer has exchanged his birthright in them for the wattage of his all-night sun. His children will never know the blessed dark of night.

–Leslie Peltier, *Starlight Nights,* Harper and Row, 1965

In 2001, NASA warned in its "Science News" web pages: "The Milky Way is dimming, not because the end of the Universe is near, but rather as a result of light pollution: the inadvertent illumination of the atmosphere from street lights, outdoor advertising, homes, schools, airports and other sources. Every night billions of bulbs send their energy skyward where microscopic bits of matter—air molecules, airborne dust, and water vapor droplets—reflect much of the wasted light back to Earth" (Fig. 8.1).

By 2009, the problem of light pollution, gradually veiling the night sky since the 1950s, had become serious—so much so that the UK's Royal Commission on Environmental Pollution investigated the subject and made recommendations to the British Government (see Appendix 3). Their report was entitled *Artificial Light in the Environment* (© Crown 2009).

© Springer International Publishing Switzerland 2016
B. Mizon, *Finding a Million-Star Hotel*, The Patrick Moore Practical Astronomy Series,
DOI 10.1007/978-3-319-33855-2_8

Fig. 8.1 Orion veiled by wasted light. (Photo: Bob Mizon)

The report begins:

Imagine a vista of outstanding natural beauty, to say nothing of historic and cultural significance, permanently obscured from public view by a cloud of non-toxic, but visually impenetrable, artificial vapor. Such a prospect seems unthinkable in Britain today. Yet we seem to tolerate the daily destruction of arguably the most culturally universal and historically pristine of natural vistas – the night sky, filled with constellations of stars, and planets and galaxies. The responsible pollutant, however, is not an impenetrable vapor, but the light that we so freely emit into our surroundings. (www.gov.uk/government/uploads/system/uploads/attachment_data/file/228832/9780108508547.pdf.pdf)

In urban areas, stray artificial light has proliferated over the decades to fill almost every corner of the night-time environment. Paolo Nespoli's photos of the world at night taken from the International Space Station illustrate this only too well (Fig. 8.2). In rural spaces, light from distant towns, sports facilities and roads color the horizon and detract from the character of the countryside at night. Unable to compete, the light from distant stars and galaxies, which has taken hundreds, thousands, even millions of years to reach our planet, is lost to us in the last milli-second of its unimaginably long journey. There are some who say that the battle against poor quality lighting is just too steep a hill to climb, and that dark sky reserves and sites are the answer. They forget that the tide of wasted light from outside these havens will one day wash over what remains of rural night skies, unless lights are properly controlled and regulated.

Stars do not belong just in dark sky zoos. We are all entitled to the optimum night sky wherever we live, even if it means seeing only the main stars in the most prominent constellations from urban sites.

The problem worsens annually. There are few places left in mainland England where a pristine night sky can be seen. In much of the United States, especially east of the Mississippi and in the populated areas of the west coast, the baleful glow from distant cities colors the night sky, even in formerly dark rural areas. Both wildlife and humans need the night. It is becoming increasingly hard to find for vast numbers of Earth's inhabitants.

It's an interesting fact that while just about all electrical devices do exactly what they are designed to do, lights often don't. Hair driers dry hair, cookers cook, irons flatten things and refrigerators refrigerate. But most of the lights we see around us at night don't just light the area to be lit; they are thoughtlessly angled, sending their emissions into nearby trees and houses, gardens and fields, and off into the countryside and the sky. The night vision of approaching drivers and pedestrians is compromised, just when they need it most, by unnecessary glare. The stars are veiled, sometimes to near invisibility. Whole generations have grown up assuming that this wasteful lighting is normal. It has been around for longer than any of us has lived. Criminals use the wall of glare created by some domestic and commer-cial exterior floodlights to go about their unwelcome business unseen by potential witnesses (Fig. 8.3).

Even those living in remote, unlit villages may not see the pristine, dark vault of the heavens, since light pollution travels. A big city may stamp its light 'signature' on the horizon from a great distance away (Fig. 8.4). Residents in a British National Park, the New Forest in south-central England, have large tracts of their night sky tainted by light from two big towns (see Chap. 3). In 2007, Southampton Council even proposed shining long-distance laser beams from their Town Hall tower across the night skies of the New Forest. The scheme was abandoned after vigorous protest from many individuals and organizations.

The U. S. National Park Service (NPS) has a Night Sky Team, formed in 1999. It documents and monitors the intrusion of light pollution in many of the country's fifty-nine national parks. Light levels are checked with cameras and meters, using robotic telescope technology. It was soon realized that even in remote and wild

Fig. 8.2 The United Kingdom at night from the International Space Station. (Photo courtesy ESA/NASA/P. Nespoli)

Fig. 8.3 Wrongly aimed domestic floodlight. Much of its output goes into the sky. (Photo: Martin Morgan-Taylor)

areas, light domes from distant inhabited areas can intrude above the horizon. The NPS Night Sky team has collected initial data on light-pollution conditions in twenty national parks in seven states. According to NPS Night Sky project manager Chad Moore (Bryce Canyon National Park, Utah): "At every park we have surveyed, we have detected artificial lights." Leakage from the lights of Las Vegas, for example, intrudes upon Death Valley and Joshua Tree National Parks in California, and is reportedly visible 200 miles (320 km) from its source. Light traveling between 0° and 10° above the horizontal causes most skyglow, scattering from countless particles and aerosols crowding its long path—a process described in detail in, for example, an article *Towards Understanding Skyglow* written by Dr. Christopher Baddiley and Tom Webster in 2007 and published by the UK's Institution of Lighting Professionals.

Sky & Telescope magazine, ever a champion of everybody's right to see stars, has praised the NPS's work, in collaboration with the IDA, to improve and preserve starry nights: "The NPS commitment to night-sky monitoring is an important step for light-pollution opponents—it demonstrates that a government organization at the federal level is taking action to protect star-filled skies as a valuable natural resource, and to correct careless lighting practices."

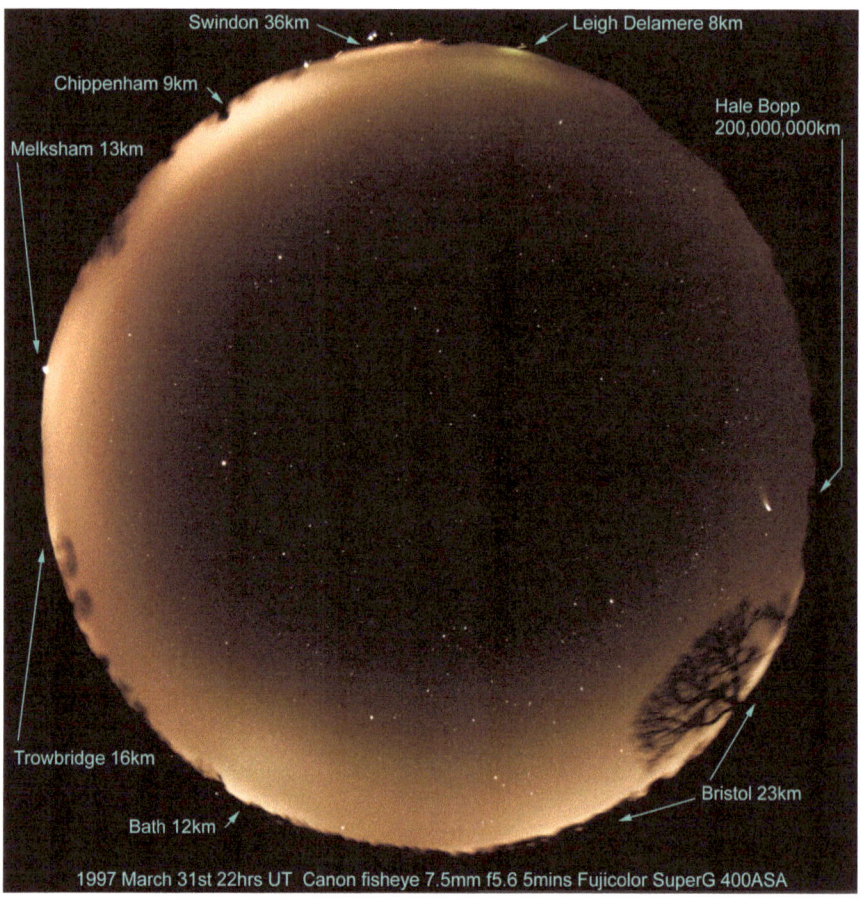

Fig. 8.4 An all-sky camera shows light pollution from distant towns leaking into a rural night sky. (Photo: Mike Tabb)

There are also many lights allowed to taint the sky from *within* rural spaces. We see the rising glow from industrial processes, farms, vehicle depots and other premises whose owners often mistakenly think that bright, glaring lights, facing outwards rather than being confined to the premises, protect their assets. Floodlit golf driving ranges cause local uproar. Even the ancient dark skies over Stonehenge, arguably western Europe's most astronomically significant ancient monument, and a UNESCO World Heritage Site, have been compromised by a driving range's floodlights (Fig. 8.5).

There is a subtle irony in the fact that most people in the developed world cannot properly see or relate to a starry sky, knowing only the baleful glow of stray light that hides it; for those hidden stars are the very forges within which we have our origins, as does everything that surrounds us. Stars have been made irrelevant in our lives.

Fig. 8.5 Effect of a golf driving range on ancient stars over Stonehenge. (Photo: CfDS image library)

If correctly designed and shielded, and not too bright for the lighting task, exterior lights can enhance the nocturnal scene, help us to see, and not waste money and energy. Most of those campaigning for a better-lit night-time environment do not want to turn off a single necessary light. What is needed to minimize light pollution is simply the right amount of light, directed where needed. Shielding, good design and sensitivity about the effects of light on others can lead to a solution.

Most exterior lighting manufacturers now produce well-directed lamps, though the race between better lighting technology and the continued proliferation of poorly aimed lights is a close one. Street lights in many countries are now rapidly being converted to types with flat-glass lenses that allow light to be emitted only downwards. What Italian dark sky campaigner Fabio Falchi called a "tsunami" of LED street lights is sweeping the developed world (Figs. 8.6 and 8.7). Good sense and decades of hard work by campaigners and thoughtful lighting professionals mean that the new LEDs are almost always downward-directed. Job done? If too bright for the lighting task, as is often the case, LEDs can reintroduce problems of stray and intrusive light and skyglow. Excessive light output means that some of their emissions will inevitably be reflected skywards from the ground. LEDs with a blue-rich spectrum also disrupt the lives of countless nocturnal species. In residential areas, they cause unhealthy light intrusion into premises.

Fig. 8.6 An LED streetlight, Isle of Wight, southern England. (Photo: Chris Wood)

The IDA states, at darksky.org/lighting/led-practical-guide, that "every effort should be made to diminish or eliminate blue-light exposure after dark."

American amateur astronomer and author Bob King, author of the widely read and respected "Astro Bob" blog, wrote of the new LED lighting in Duluth: "Buildings and intersections that had been orange the night before were bathed in a far more intense blue-white light… Our city engineers deserve high marks for adhering to good lighting standards by packaging the new lights in shielded housings with minimal light spill upwards and to the sides." Any authority putting in lights that shine above the horizontal nowadays really is, from a lighting professional's viewpoint, living in the past. However, Bob added, "Still, everything was not OK. The LED street lights were INTENSELY bright, much more so than the 'old-fashioned' sodiums."

So even well-directed modern lights can taint the night sky through excessive output, leading to upward reflections.

Amateur astronomers and others who simply like seeing stars and object to upward light spill should not be treated by officialdom as special, hypersensitive cases with an unusual minority pastime. In the Georgian city of Bath, England, in February 2006, a UK government legislation seminar examined the recently introduced Clean Neighborhoods and Environment Act (2005). Its section 102 is entitled "Artificial Lighting as a Statutory Nuisance." This was the first UK law covering light nuisance, but it contained nothing about the protection of the night

Fig. 8.7 An LED streetlight in Hawaii. (Photo: Darren Baskill)

sky; its remit was light entering premises. At this seminar, a representative of DEFRA, the UK government's environment department, stated that "It should not be the case that astronomers are exempted from the legislation," and "they should not be considered as sensitive complainants." A desire to see the night sky is not an inexplicable thing, though some administrators, presumably those who live in areas beneath tainted skies, seem to think so.

On June 16, 2009, the American Medical Association (AMA) voted unanimously to support efforts to control light pollution. Why has the AMA, a wide-ranging and influential group, decided to support light-pollution legislation? They cite glare from bad lighting as a health hazard, unnecessary energy waste, extra CO_2 produced, and finally, the fact that all species (including humans) need darkness to survive and thrive. Nearly every organism on Earth, with the exception of some abyssal sea creatures and species that have evolved to live permanently in the darkness of caves, has wired into its physiology a set of responses to the day-night cycle caused by our planet's rotation. We tamper with this ancient programming at our peril.

Efficiently lit streets, more stars—a view of the night sky and a well-lit environment are not incompatible. If you can't see the stars because of neighbors' intrusive lighting, which can be even more disturbing than local 'lawn-sprinkler' street lighting, use the websites[1] of the IDA and the CfDS to prepare a reasoned and positive

---

[1] www.darksky.org/outdoorlighting/guidance; www.britastro.org/dark-skies inf007.htm?2O.

Fig. 8.8 The glittering curve of the constellation of Perseus. (Photo: James Hilder)

approach, showing them why things should not be so. If neighbors can't see reason, you may have to turn to local nuisance laws.

Ask those installing new street lighting if it will be of the modern variety, with flat-glass lenses and proper shielding. Lone voices are often, it seems, too faint to hear within the corridors of power, so ask other residents to add their views to yours if they think that lighting could be improved.

Screening off any particularly intrusive lights can work well, and showing neighbors the night sky through a telescope or binoculars may make them think about their lighting strategy. A neighbor once looked through the author's telescope at the Double Cluster, two adjacent masses of stars in the summer constellation of Perseus. This glittering double crowd of stars can be seen as a tight condensation at the top center of James Hilder's striking photo (Fig. 8.8) of the Perseus Milky Way, taken at Galloway Forest Park. The Double Cluster is 7,500 light years away, and unglamorously labeled NGC 869 and NGC 884. My neighbor compared it to a spilled jewel box in the sky. Not long afterwards, I noticed that her outdoor light had been re-angled and the light source changed to a lower wattage.

The nitrogen in our DNA, the calcium in our teeth, the iron in our blood, the carbon in our apple pies were made in the interiors of collapsing stars. We are made of starstuff.

–Carl Sagan, *Cosmos*

The starry sky deserves as much respect as the rest of our environment, since ancient stars literally made most of our component atoms. They are our chemical parents.

It's a truism that the various elements that make up the human body and are vital to its functioning can be reduced to a large puddle of water, an assortment of iron nails, matchheads and other basic stuff. Few ever wonder where these elements originate. Astronomers love to reveal the fact that we're made from the nuclear ash of dead stars.

The matter in our flesh and bones, the various elements that facilitate the electrical activity in our brains and the functioning of our organs—all were cooked up, built from lighter atoms, in the interiors of old, collapsing or exploding stars billions of years ago. We contain traces of quite unexpected starstuff: arsenic, gold, molybdenum, vanadium, and so on.

Many of those element-forging stars are now white dwarfs, invisible to the largest telescopes, swept through the distant reaches of our giant Milky Way Galaxy by its majestic 250-million-year rotation. Earth and its trillions of life-forms travel at 514,000 mph (143 mi/s), or 828,000 km/h (230 km/s), on their journey around the center of the galaxy.

Further back still, at the very beginning, all the hydrogen that now makes up most of the universe—and, as a component of water, a significant proportion of the bodies of living creatures—was formed in the Big Bang, the stupendous explosion that kick-started the cosmos as we know it today. We still detect the faint background hiss that is the echo of this event of 14 billion years ago.

Whatever is left of our long-departed bodies in, on or around this planet when it dies, irradiated by the senile, bloated Sun a few billion years hence, will be redistributed into the clouds that swirl around our galaxy. Indeed, most of our atoms may already have been recycled through planetary systems more than once, and some may even have been in other organisms on distant planets in an unimaginably remote past.

To many people, the stars in the sky remain a symbol of remoteness, and distant astronomical objects may seem irrelevant to our lives. How does the cosmos outside Earth touch our everyday existence? We are beginning to realize that the Moon and stars, including the star only 93 million miles from us that powers planet Earth, have a lot more relevance in our lives than we ever imagined.

• The Sun, our nearest star (Fig. 8.9), keeps us alive every second of the day and night by growing our food, and by warming and circulating air so that we do not freeze to death after sunset. It provides fuel for our motive power and electricity for our appliances through energy stored for millions of years in fossil fuels: coal, oil and gas. It causes all the weather on our planet, regulates our daily biological cycles, and its presence or absence can make or break a holiday.

Fig. 8.9 The Sun, provider of our very existence. (Photo: Sheri Lynn Karl)

- Although the atoms in our bodies were made either in the Big Bang (hydrogen) or, later, in the interiors of stars, atoms heavier than iron, for example gold and platinum, are made in the immense ovens of supernovae, giant stars exploding at the end of their lives (Fig. 8.10). Because supernovae are relatively rare, the heavy elements they add to the cosmos are rare, and therefore expensive and sought after. The relative values of the commodities around us are determined by their stellar origins. People have worked, struggled, stolen, fought and died in their quest for gold, silver and other precious metals, but they wouldn't give a second glance to similar amounts of iron or carbon—all because of the relative abundances of different types of stars.
- The vast tides created by the Moon, much closer millions of years ago, drew minerals into the oceans, hastening the onset of life. The Moon was instrumental in helping living creatures establish themselves on land, where they could evolve to live on beaches and around rock pools in an amphibian stage. Tides provided a wet-dry area where they could learn to live in this mode. If Earth had had no Moon, would life ever have been able to leave the oceans? Would we all be intelligent fish?

Fig. 8.10 A supernova erupts in galaxy M95, 38 million light years away in Leo. The supernova appears as a bright point of light above the center of the galaxy. (Photo: Nick Hart)

For our hunter-gatherer human ancestors, moonlight may have meant the difference between living or dying, as it gave them extra foraging opportunities at night. Many creatures regulate their behavior, especially in breeding and egg-laying, according to the full Moon (Fig. 8.11), and the human menstrual cycle's mimicking of the Moon's cycle may be no coincidence. Our only natural satellite, especially when at its full phase, does seem to have an effect upon certain susceptible people (hence "lunacy"), though the causes are little understood.

- The genetic mutations that drive evolution are in part caused by the impacts of swift cosmic rays from space. These originate in supernova explosions and other high-energy events, so the progress of life on Earth is driven partly by the deaths of distant supermassive stars.
- On a more everyday note, our clocks and calendars are set by the motions of heavenly bodies. The rotation of Earth and the position of the Sun set the time. When the Sun is at its highest over your time zone, it is midday. Months are "moonths," time periods based on the changing phases of the Moon. Why are there 24 hours in a day, and 60 seconds in a minute? The ancient Babylonians had a base-60 counting system, and divided the sky, time and angular surfaces accordingly—hence the twelve starry houses of the zodiac through which the

Fig. 8.11 The Moon's influence on the birth of life on Earth may be greater than we think. (Photo: Shaun Reynolds)

seven bright lights in the sky move in the course of a year. There are seven days in a week, named for the seven bright objects, gods and goddesses of old, which move around in the sky against the background of distant stars: Sun (Sunday), Moon (Monday), Mars (French: *mardi*, Tuesday), Mercury (French: *mercredi*, Wednesday), Jupiter (French: *jeudi*, Thursday), Venus (French: *vendredi*, Friday), and Saturn (Saturday). In English, Norse gods or their Anglo-Saxon equivalents have usurped the days from their Roman counterparts for Tuesday (Tiw's day), Wednesday (Woden's day), Thursday (Thor's day) and Friday (Freya's day). When is Easter Sunday? It's the first Sunday after the first full Moon following the vernal equinox. This equinox is the moment in spring when the Sun crosses the equator on the sky on its apparent annual journey through the stars. The visibility of the thin crescent Moon marks the beginning of the Muslim lunar month.

• Last, but by no means least, is the part the starry night sky has played in human culture, science and religion. Astronomy, developing alongside the sky-religion of astrology, was already millennia old when physics and mathematics flowered; in its modern form, it embraces most other scientific disciplines, and overflows into history, philosophy and theology. The apparently unchanging cosmos was long ago compared with the corruptible and ever-changing Earth below, and appropriate conclusions were drawn about the perfection of the gods and the fallibility of humankind.

Fig. 8.12 Unlike those who pollute the night sky, the people who did this can be punished by the law. (Photo: Bob Mizon)

One of the saddest manifestations of the damage human technical progress can do to the environment, and indirectly to the human spirit, is the slow eradication of the visibility of the night sky, of the stars that made and inspired us. In nearly every country in the developed world, the night sky is the only part of our environment with no protection in law. It is a sad fact that, in most countries, penalties for dumping rubbish and despoiling streets, fields and roadsides with litter can be severe; polluting rivers and fouling the sea are criminal offences. But masking the beauty of the night sky with wasted light carries no penalty at all in law (Fig. 8.12).

> Half of our environment has no protection in law, and the current relighting of the world with ever brighter blue-white lights threatens it further. The current upsurge of interest in astronomy should be encouraged – if we think that this planet is all that exists we get a very wrong view of our place in the scheme of things.
>
> –Bob Mizon, Cranborne Chase dark sky seminar, November 2014

Appendix A

IDA International
Dark Sky Places
(as of April 2016)

Aoraki Mackenzie International Dark-Sky Reserve, South Island, New Zealand
Brecon Beacons National Park, Wales
Exmoor National Park, England
Kerry International Dark-Sky Reserve, Ireland
Mont-Mégantic, Québec, Canada
Namib Rand Nature Reserve, Namibia
Pic du Midi, France
Westhavelland Nature Park, Germany

Big Bend National Park, Texas
Blue Ridge Observatory and Star Park, North Carolina
Capitol Reef National Park, Utah
Chaco Culture National Historical Park, New Mexico
Cherry Springs State Park, Pennsylvania
Clayton Lake State Park, New Mexico
Copper Breaks State Park, Texas
De Boschplaat, Netherlands
Death Valley National Park, California

© Springer International Publishing Switzerland 2016
B. Mizon, *Finding a Million-Star Hotel,* The Patrick Moore Practical Astronomy Series,
DOI 10.1007/978-3-319-33855-2

Eifel National Park, Germany
Elan Valley Estate, Wales
Enchanted Rock State Natural Area, Texas
Galloway Forest Park, Scotland
Geauga Observatory ParkGrand Canyon-Parashant National Monument
Goldendale Observatory State Park, Washington
The Headlands, Michigan
Hortobágy National Park, Hungary
Hovenweep National Monument, Utah and Colorado Observatory Park, Ohio
Mayland Community College Blue Ridge Observatory and Star Park
Natural Bridges National Monument, Utah
Northumberland National Park/Kielder Forest, England
Oracle State Park, Arizona
Staunton River State Park, Virginia
Weber County North Fork Park, Utah
Zselic National Landscape Protection Area, Hungary

Beverly Shores, Indiana
Bon Accord, Canada
Borrego Springs, California
Coll, Scotland
Dripping Springs, Texas
Flagstaff, Arizona, United States
Homer Glen, Illinois
Horseshoe Bay, Texas
Sark, Channel Islands
Sedona, Arizona
Thunder Mountain Pootsee Nightsky, Arizona
Westcliffe/Silver Cliff, Colorado
For an update see www.darksky.org/night-sky-conservation/dark-sky-places.

There are many places in the UK that the Dark-Sky Discovery Scheme has designated as Dark-Sky Discovery sites. For full information and a map, see www.darkskydiscovery.org.uk.

Appendix B

UK Dark Sky
Discovery Sites

The UK Dark Sky Discovery Partnership, a publicly funded body, consists of a number of collaborating UK astronomical and environmental organizations, at both local and national levels. Its stated aims are to:

- engage people from diverse backgrounds with the night sky;
- encourage positive attitudes towards science and technology;
- support the development of dark sky places, awareness and tourism
- develop a national network of dark sky communicators;
- create long-lasting organizational partnerships in this area.

The national DSD Partnership is led by the Science and Technology Facilities Council (STFC). Its other members are the British Association of Planetaria, the Commission for Dark Skies, the Federation of Astronomical Societies, the Institute of Physics, the Royal Astronomical Society and the Society for Popular Astronomy.

There are many places in the UK the Dark Sky Discovery Scheme has designated as Dark Sky Discovery sites.

For more information and an interactive map, see www.darkskydiscovery.org.uk.

© Springer International Publishing Switzerland 2016   285
B. Mizon, *Finding a Million-Star Hotel,* The Patrick Moore Practical Astronomy Series,
DOI 10.1007/978-3-319-33855-2

Appendix C

The Report of the Royal Commission on Environmental Pollution

Following are excerpts from the Report of the Royal Commission on Environmental Pollution, *Artificial Light in the Environment*, 2009, its conclusions and recommendations.

We have examined the explosive growth in outdoor lighting in the UK since the Second World War and the resulting loss of visual amenity of the night sky due to light pollution, particularly sky glow. We believe that access to the natural beauty of the night sky is every bit as important as the preservation of other aspects of natural beauty which society routinely seeks to protect for the enjoyment of its citizens and for posterity.

We are also concerned that we simply do not know enough about the biological impacts of light pollution on plants and wildlife, particularly at the population and ecosystem levels. In many cases scientists have barely begun to look. Humans, and most other animals and plants, have evolved in an environment which has alternating periods of light and darkness, both within each day and, outside the tropics, between seasons. This has important consequences for the ways in which organisms behave in the environment and for certain processes within organisms and ecosystems.

The disruption of normal light patterns can therefore have significant effects.

© Springer International Publishing Switzerland 2016
B. Mizon, *Finding a Million-Star Hotel,* The Patrick Moore Practical Astronomy Series,
DOI 10.1007/978-3-319-33855-2

However, because of the general perception of light as a natural and benign phenomenon, it is sometimes difficult for people to understand its negative effects. This may be one reason for the apparent general indifference to the potential negative impacts of light in the environment when any other anthropogenic effect having the same impact would have long been subject to more rigorous scrutiny and control.

To rapidly redress the lack of access to the night sky for the population of the UK, we recommend that those responsible for the management of existing National Parks and Areas of Outstanding Natural Beauty and the equivalent National Scenic Areas in Scotland seek to eliminate unnecessary outdoor light and to better design and manage that which cannot be eliminated, and also that efforts are made to retain or create dark skies over urban areas so that people in major centers of population may have access to the night sky.

Much light pollution comes from road lighting. While we recognize that road lights can reduce road accidents and increase people's sense of security, we have found that the magnitude of claimed benefits are either smaller than previously thought (accidents) or equivocal (crime reduction), and in any case these goals are not necessarily associated with higher levels of illumination. Smarter lighting rather than more lighting is the key. We recognize that lighting arrangements can be important in providing people with a sense of place, but inappropriate lighting causes unnecessary stress to people, to plants and to animals. We recommend that the highway authorities and local authorities reassess the lighting of roads against potential road safety and crime reduction benefits.

Private lighting of external space is a growing cause for concern. We recommend that the sale of all new external lighting and floodlighting is accompanied by best practice advice, in order to help installers to aim them correctly, so as to avoid light nuisance and minimize light pollution.

Light is one of the factors covered in Government Planning Policy Statements, and the impact of artificial light from developments has to be investigated in the preparation of an environmental statement under Environmental Impact Assessment Regulations. We recommend that there should be explicit consideration of light in planning policy. We recommend that planning guidance includes a presumption against the provision of artificial light in some areas where it may have a negative impact on species of concern. We also recommend that guidance is expanded specifically to enable local authorities to assess the likely ecological impacts of changes to the amount and quality of artificial light. Similar guidance should be provided by the Devolved Administrations.

Because we consider that more explicit recognition needs to be given to the visual and wider societal impacts of artificial lighting, particularly in urban areas, we recommend that local authorities should develop a lighting master plan in consultation with their local communities, professional lighting designers, and their own public lighting engineers.

We recognize that lighting of public space has benefits and is perceived by many in society to have advantages which outweigh the negative impacts. But it is equally clear that improved technology can deliver those benefits while minimizing unwanted side-effects. The technologies to do this already exist. We recommend that lighting standards should require the provision of light at an intensity no greater than the minimum necessary to deliver the intended benefits and that the light should be directed at only those areas which are intended to be illuminated.

Finally, at present, none of these issues appear to have any natural locus within Government, with different departments considering different aspects and some key departments such as transport explicitly not considering impacts on the natural environment. Light has been the poor relation for too long; Government needs to accept the fact that light, like noise and chemicals, in the wrong quantity, in the wrong place and at the wrong time can cause problems and must be addressed explicitly in policy development. We recommend that DEFRA and equivalent bodies elsewhere in the UK take the lead in coordinating interdepartmental activity on artificial light.

In closing, we emphasize that while research into and monitoring of the biological effects of light pollution on human wellbeing and natural ecosystems are desirable, this may not be an issue which requires greater scientific confidence to justify corrective action. We are convinced of the plausibility of the argument that light in the wrong place and at the wrong time can disturb the lives of organisms, potentially with adverse ecological effects. Considered alongside the indisputable loss of the visual amenity of the night sky, the Royal Commission considers that there is sufficient reason to take action to reduce light pollution without waiting for the results of scientific inquiry into biological impacts, which should focus on research that could inform the design of new lighting technologies and installation practices.

The full report is at www.rcep.org.uk/reports/sr-2009-light/sr-light.htm.

Appendix D

Starry Starry Night

In 1993, the British Astronomical Association's Commission for Dark Skies (CfDS) joined forces with the Campaign to Protect Rural England (CPRE) to produce the widely distributed and much quoted booklet *Starry Starry Night*, extensively revised in 2010 and 2013. Its contents may be of use to anyone seeking to combat both urban and rural light waste, and the bulk of its text is reproduced here:

Human beings have long looked up in awe, on cloudless nights, at the star-strewn heavens. What did our distant ancestors make of it all? They drew the stars into the framework of their lives by creating constellations, fitting them to their beliefs and myths. They marveled at the ghostly river of light which is the Milky Way, our own galaxy of 200 billion stars seen from within, arching across the sky. The stars, the moving planets, and ephemeral events such as aurorae, comets and meteors, a—all these have inspired religious beliefs, poetry, music and scientific enquiry. The mysterious and unreachable vault of the heavens has been a primary stimulus to the human faculties of wonder and discovery.

For countless years, all this has been ours on every clear night. But during the twentieth century, the glory of the night sky was quietly and gradually taken away from most of the world's people by wasted artificial light. This process continues unabated, and at a rapidly accelerating rate. Satellite images of Earth at night show wasted light energy from every town and city, along roads, and in rural areas. Even in the countryside, poorly aimed, over-bright floodlights and security lamps have stolen the blessed night from humans, and countless other species that have evolved to the rhythm of light and darkness. The day-night cycles, behavior, feeding and mating patterns of bats, birds, glow-worms, moths, and countless other species are disturbed, and millions are killed, by light going where it is not needed.

© Springer International Publishing Switzerland 2016
B. Mizon, *Finding a Million-Star Hotel,* The Patrick Moore Practical Astronomy Series,
DOI 10.1007/978-3-319-33855-2

Why Can We See Stars? The Threat to Our Skies

The light we see in the night sky is mostly direct spillage from lamps that have simply not been designed for the lighting task. Their emissions trespass into neighboring areas, and into the sky. They will often be too bright, which adds to glare and skyglow. Light traveling upwards is scattered and reflected by ever-present tiny particles and water droplets in the air, even on the clearest nights. The result is a baleful glow in the night sky, now seen from nearly everywhere in the UK. The constellations, aurorae, meteors and the zodiacal light, the faint reflection from billions of dust particles in the plane of the Solar System, are now things of the past for many of us.

There is no doubt that the spread of public lighting since the mid-1800s has brought great benefits. The quality and efficiency of lamps are continually improved, but the 'poor relation' in lighting design is directionality. What the Victorians saw as a blessing has become an environmental blight. Glare, over-lighting and skyglow have tainted the night. Light intrusion into others' premises is now a major cause of complaints to environmental health officers, and research suggests it is damaging to health. The Clean Neighborhoods and Environment Act of 2005 gives local authorities powers to address intrusive light nuisance, but the night sky itself still has no real protection in law.

The stars continue to disappear behind the veil of wasted light, over great cities and smaller towns. Bedrooms are filled with light even with curtains closed. Aggressive 500-watt floodlights turn neighbor against neighbor in both town and countryside, ousting the traditional more modest and welcoming porch light. It is an interesting fact that Britain's brightest lighthouse, the Longstone on the Farne Islands, has a 1000-watt source, yet many of us, even those who pay lip service to protecting the environment, use half this amount to light our gardens and drives. Another interesting fact: a 100-watt bulb left on all night for one year releases a quarter of a ton of carbon dioxide, the major greenhouse gas, from the burning of the fossil fuel used to power it. Little protest is made about wasted light. Is light pollution merely the uncomfortable cost of progress?

Unlike many other forms of pollution, light pollution is fairly easily reversible. Lights can be shielded or replaced with more appropriate designs, and wattages can be adjusted appropriately. In the words of the Institution of Lighting Professionals: "Light pollution, whether it keeps you awake through a bedroom window or impedes your view of the night sky, is a form of pollution and could be substantially reduced without detriment to the lighting task."

Can we regain our heritage above? Yes. Visit the websites listed on www.britastro.org/dark-skies.

Types of Lighting

The earliest practical lights were incandescent tungsten bulbs, still used commonly for domestic purposes. The next development in street lighting was mercury-vapor discharge lamps, which give a blue-white light, but are low efficiency and have a fairly short life.

Many of the glary, over-powered "security" lamps sold nowadays are of the tungsten-halogen type, very inefficient and short-lived. Then came low-pressure sodium (SOX), the strong orange light beneath which colors are indistinguishable. It is the most energy-efficient lighting, and the lamps have a long life. High-pressure sodium (SON) started to replace SOX in the late 1970s. The SON lamps are much smaller than SOX, which can be nearly a meter long. SON can therefore be more easily enclosed in a reflector that directs the light where it is needed. SON energy efficiency is not as good as SOX, but the life is even better. Today there is a move towards smaller lamps, for example halogen, with white light, and better color rendition, though the lifetime may be shorter.

Long-lived light-emitting diodes (LEDs) are beginning to appear above our streets, though sadly far too many of them are very bright—too bright for the lighting task—and of a blue-rich spectrum, which is bad news for wildlife and for humans trying to get to sleep.

It can help to talk to people about the skyglow issue, stressing energy and money wasted. What would they think if water mains leaked every few meters? The CfDS does not want to switch off any necessary light; its motto is "the right amount of light, directed where needed."

Are your local media up to date with the skyglow issue? Do they include skyglow in their environmental reporting?

Ask neighbors about lighting plans and tell them why you enjoy the night sky. Politely approach owners of obtrusive lights. They may not know they are causing a problem. Experience shows that most offenders will take some remedial action. Write to local councilors, council lighting/highway engineers, MPs, MEPs, sports clubs, etc., to ask about their views and lighting policies.

If new, less glary lighting is perceived by some to be dimmer, make sure that they understand the efficiency of modern, better-directed lamps. Not seeing the glare is a good thing. Set a good example by not using over-bright and glary exterior lights on your own premises.

We are told that the climate and the environment in general are under threat from energy waste. Ensure that debate in your area recognizes the contribution that light spillage makes to these problems. Remember that 100-watt bulb?

If you or any group you belong to has a website, link to the Commission for Dark Skies on www.britastro.org/dark-skies.

Try to forestall poor lighting schemes by studying planning applications and making sure your council has lighting clauses in its planning and environmental strategies. Help CfDS directly by subscribing to its newsletter, donating to its fighting fund, becoming a local officer or distributing its literature.

Remember: 'broadsides,' carping criticism and baldly accusing someone of being a polluter are counterproductive strategies. We can reclaim the night sky through reasoned argument and strength in numbers. Nothing positive comes from light pollution. Everyone wins if it is reduced.

The British Astronomical Association's Commission for Dark Skies and other such groups work to ensure star-quality lighting. Their networks publicize the problem, praise good practice and strive to turn poor lighting schemes into more acceptable ones.

Appendix E

100 Places
in the United States
to See the August 2017
Total Eclipse
of the Sun

These places, some large and some small, are listed in order from west to east as the Moon's shadow encounters them. Those in **bold** are within the track of the shadow. Others are very near the edge of the track. Those described as being on the center line are either on it or near enough to it to receive the maximum possible duration of the eclipse as experienced from that point.

Blue Ridge, Atlanta, **Helen**, **Dillard (on center line)**, Athens, Augusta.

Nampa, Boise, **Boise National Forest (part)**, **Sawtooth National Forest (part)**, **Stanley**, Twin Falls, **Sun Valley**, Pocatello, **Terreton (on center line)**, **Idaho Falls**, **Rexburg**.

Junction City, Holton, Topeka, **Reserve (on centre line)**, Lawrence, **Atchison**.

© Springer International Publishing Switzerland 2016
B. Mizon, *Finding a Million-Star Hotel,* The Patrick Moore Practical Astronomy Series,
DOI 10.1007/978-3-319-33855-2

Kentucky

Hopkinsville (on center line), Madisonville, Evansville, Bowling Green.

Missouri

St Joseph (on center line), Kansas City, Marshall (on center line), Columbia, Lebanon, Jefferson City, Farmington, St Louis, Cape Girardeau, Carbondale, Mount Vernon.

Nebraska

Scottsbluff, Chadron, Alliance, North Platte, Ravenna (on centre line), Hastings, Kearney, Grand Island (on centre line), Columbus, Lincoln, Filley (on center line), Omaha.

North Carolina

Nantahala National Forest (part), Andrews (on center line), Franklin, Cherokee, Brevard, Asheville, Hendersonville.

Oregon

Newport, Lincoln City, Corvallis, Albany, Salem, Lyons (on center line), Mount Hood National Forest (part), Eugene, Portland, Bend, Redmond, Pendleton, La Grande, Baker City.

South Carolina

Greenville, Spartanburg, Sumter National Forest, Columbia, Orangeburg, Sumter, Bonneau (on center line), Charleston, McClellanville, Georgetown, Myrtle Beach.

Tennessee

Clarksville, Nashville, Murfreesboro, Gordonsville (on center line), Chattanooga, Knoxville.

Jackson, Grand Teton National Park, Yellowstone National Park (part)/ **Shoshone National Forest (part)**, **Dubois**, **Pavillion (on center line)**, **Riverton**, **Thermopolis**, **Casper (on center line)**, **Douglas**, Cheyenne, **Fort Laramie.**

See also:

www.eclipse2017.org/2017/in_the_path.htm

www.exploratorium.edu/eclipse/how.html

Appendix F

Further Information

www.astronomyforum.net/astronomy-locations/usa
www.astunit.com
www.binocularsky.com
www.heavens-above.com
www.skymaps.com
www.skyandtelescope.com
www.spaceweather.com

Lists of astronomical societies in the UK and the United States that might give advice on good observing sites may be found at:

UK: www.astronomyclubs.co.uk/Clubs/Counties.aspx
USA: www.astroleague.org/al/general/society.html

General

Dickinson, T. Dyer, A. *The Backyard Astronomer's Guide* (Firefly Books, 2008).
Karkoschka, E. *Observer's Sky Atlas* (Springer, 2007).
Sinnott, R. *Sky & Telescope's Pocket Sky Atlas, Jumbo Edition* (Sky, 2016).

© Springer International Publishing Switzerland 2016   299
B. Mizon, *Finding a Million-Star Hotel,* The Patrick Moore Practical Astronomy Series,
DOI 10.1007/978-3-319-33855-2

Tonkin, S. *Binocular Astronomy* (Springer, The Patrick Moore Practical Astronomy Series, 2nd edition, 2013).
Norton's Star Atlas and Reference Handbook (ed. Ian Ridpath, Dutton, 2003).

Astronomical Equipment

English, N. *Choosing and Using a Refracting Telescope* (Springer, 2010)
Pugh, P. *The Science and Art of Using Telescopes* (Springer, 2009)
Philip's Planisphere (G. Philip & Son, 2012)
downloads.bbc.co.uk/tv/stargazinglive/sgl_guide_to_telescopes.pdf
www.skyandtelescope.com/astronomy-equipment/telescope-buying-guide

Light Pollution

Baddiley, Dr. C. J. *Towards Understanding Skyglow* (Available through CfDS and ILP, 2007)
   Bogard, P. *The End of Night* (Little, Brown and Company, 2013)
   Cinzano, P., Falchi, F., and Elvidge, D. *The First World Atlas of Artificial Night Sky Brightness* (R.A.S., 2001)
   Commission for Dark Skies. *Blinded by the Light? A Handbook on Light Pollution* (Order via www.britastro.org/dark-skies, 2013).
   FLAP. Fatal Light Awareness Program, Toronto. (Various publications. www.flap.org).
   Klinkenborg, V. *Our Vanishing Night* (National Geographic, Nov 2008).
   Lockley, S. *Human health implications of light pollution* in *Blinded by the Light?* (CfDS, 2009).
   Mizon, B. *Light Pollution: Responses and Remedies (2nd edition)* (Springer, 2012).
   Mizon, B., and Baskill, D. *Stolen Stars: a young adult's guide to light pollution* (www.britastro.org/dark-skies, 2010).
   Morgan-Taylor, M. *"And God Divided the Light From the Darkness": Has Humanity Mixed Them Up Again?* Environmental Law and Management, (January-February, 1997).
   Rich, C., and Longcore, T. *Ecological consequences of artificial night lighting* (Island Press, 2006).
   Tonkin, S. *Astro FAQs* (Springer, 2000).
   A long list of articles on light pollution can be found on:
   www.weasner.com/etx/lp

Astronomy (USA)
Sky and Telescope (USA)
Astronomy Now (UK)
Sky at Night (UK)

For information on downloading these and others, see freeware.intrastar.net/plane-
    tarium.htm and en.wikipedia.org/wiki/Planetarium_software
Cartes du Ciel
Distant Suns
Encyclopaedia Galactica
Home Planet
Night Vision
Sky Atlas
Skyorb
Solar System 3D Simulator
Starmap (basic)
Starry Night Sky
Stellarium (a favorite)
Tuba (binocular users)

www.skymaps.com
freestarcharts.com

Appendix G

The Bortle Scale
and SQM Readings

The table below compares Sky Quality Meter values (see Chap. 1) with Bortle scale values, and gives the limiting magnitude (faintest star seen). The limiting magnitude figures assume good eyesight, and in the case of Bortle 1 and 2, exceptional vision.

SQM	Bortle scale	Limiting magnitude
22.00-21.99	1	8
21.99-21.89	2	7.1
21.89-21.69	3	6.6
21.69-21.25	4	6.2
21.25-19.50	4.5–5	5.6
19.50-18.95	6	5.0
18.95-18.38	7	4.5
18.38-17.80	8	4.0
<17.80	9	<4.0

© Springer International Publishing Switzerland 2016
B. Mizon, *Finding a Million-Star Hotel*, The Patrick Moore Practical Astronomy Series,
DOI 10.1007/978-3-319-33855-2

Appendix H

Glossary of
Astronomy Terms

This is a list of astronomy words and phrases that may require further explanation. A separate glossary of terms used in discussion of light pollution issues follows.

Airglow   Airglow can be seen in the night sky from very dark places on moonless nights. It is a faint glow in the sky created by energy within our planet's atmosphere, so in fact the sky at night would not ever be totally dark, even if starlight and diffused light from the Sun below the horizon were removed. Airglow is also referred to as nightglow.

AONB   In England, Wales and Northern Ireland certain tracts of the countryside with significant landscapes, some many hundreds of square miles in size, have been designated for special conservation. These are known as Areas of Outstanding Natural Beauty (AONBs). There are 46 of them in the United Kingdom. There are currently 33 AONBs in England, four in Wales, one straddling the border between England and Wales, and eight in Northern Ireland. Altogether, they occupy nearly 20% of the UK's rural areas.

Apogee   The Moon is said to be at apogee when it is at its furthest point from Earth.

Asteroid   Rocky bodies once commonly known as minor planets, asteroids orbit the Sun at various distances from the Sun within the Solar System, but most are found orbiting in the region between Mars and Jupiter. Eighteenth-century astronomer Sir William Herschel came up with the term *asteroid*, meaning "starlike," since through a telescope that is exactly what asteroids look like—faint points of light that can be differentiated from the starry background only because, if observed for a period of time they can be seen to be slowly moving while the stars remain 'fixed.'

© Springer International Publishing Switzerland 2016   305
B. Mizon, *Finding a Million-Star Hotel*, The Patrick Moore Practical Astronomy Series,
DOI 10.1007/978-3-319-33855-2

The first asteroid to be discovered, Ceres, was found on the first day of the nineteenth century (January 1, 1801) by Giuseppe Piazzi in Sicily (Ceres was the patron goddess of Sicily), and it was at first thought to be a new planet, but when many other similar bodies began to be discovered as the century wore on, it was realized that it was one of a class of small objects orbiting the Sun. Ceres, about 900 km in diameter, became a dwarf planet in 2006, and Vesta is now be considered the largest asteroid, at about 550 km; an example of a very small asteroid is 1991 BA, about 10 m across, which passed within 160,000 km (100,000 miles) of Earth in January 1991.

Astronomical twilight   This is the point when the Sun is 18 degrees below the observer's horizon. It is defined as the time on the clock when the light from the Sun is no longer apparent in the sky, and it becomes possible to see fainter stars.

Astrophotography   Photography of the night sky, or the Sun. Staggering photos of our universe have been taken by amateur astronomers (and some are in this book), but, contrary to what many believe, astrophotography does not have to be an arcane art backed up by a knowledge of computer processing and expensive equipment. It is quite easy to take photographs of the night sky through small, everyday digital cameras. Seek advice when photographing the Sun. A good source of information is www.skyandtelescope.com/astronomy-resources/astrophotography-tips.

Astro-tourism   A recently coined phrase covering the growing trend for tourists to seek out views of the night sky. Astro-tourism is now a significant sector in the world's tourism industry.

Auroral oval   When aurorae (northern lights) form, they settle as curtain-like structures in a band (the auroral zone) at latitudes around 70°–80° N. This band, or oval, is about 3° to 6° wide but can be seen from nearby latitudes, too, in a clear dark sky. When the Sun has sent particularly energetic material towards Earth, the oval can grow and may sink down across the planet. On March 13, 1989, for example, aurorae were seen across the world, and the magnificent display, seen even from the equator, was evidence of a geomagnetic storm that had severe consequences for satellites and the planet's electricity grids.

Averted vision   Because of the structure of the human eye, which has more of its night-sensitive cells away from its central axis, astronomers use averted vision for viewing faint objects. If you do not look directly at an object, with either the unaided eye or through optical instruments, but a little more to the side of it, you will be able to see it better.

Binocular   Traditionally, two small telescopic tubes mounted together were a 'binocular,' just as one small telescopic tube is still called a monocular. Binoculars have been used since the seventeenth century, but just when a binocular became 'a pair of binoculars' in everyday conversation seems unknown.

Black hole   Black holes, gravity 'wells' that are so dense that not even light can escape (hence 'black'), are now thought to be present in galaxies all over the universe. Some may have formed in the very early universe. If an enormous star many times more massive than the Sun collapses in upon itself, a black hole may result. Many galaxies have supermassive black holes at their cores.

Celestial equator   The projection of Earth's equator onto the sky. Because of Earth's 23.4° axial tilt, the celestial equator is inclined to the ecliptic, the plane of the Sun's apparent annual motion through the sky, by the same angle. At Earth's equator, the celestial equator passes above your head, through the zenith, in the sky. At the poles, it is theoretically along the horizon.

Cinque Port   The Confederation of Cinque Ports is a set of English coastal towns in Kent and Sussex, originally confederated for trading and military purposes, now entirely ceremonial in nature. The name is from the French *cinq* (five), and the original Cinque Ports were Hastings, New Romney, Hythe, Dover and Sandwich. New Romney was silted up and Rye became a Cinque Port.

Circumpolar   This denotes a constellation or star that never sets below the horizon for a given observer. For example, from London or southern Canada, Capella is circumpolar, skirting the horizon as Earth's rotation causes it to move anti-clockwise around the North Pole of the sky.

Comet   Relatively primitive, small Solar System bodies (and they undoubtedly exist in exoplanetary systems, too), comets consist of a nucleus of ices, rocky material and dust. Many have enormously elongated orbits and visit the Sun with periods of sometimes thousands of years. As they approach the Sun, outgassing occurs, and a 'tail' of gas and dust particles forms, pointing away from the Sun. Comets have low gravity; for example, the Philae lander sitting on Comet Churyomov-Gerasimenko (67P, target of the Rosetta mission) weighed 100 kg while on Earth, but weighs one gram on the comet.

Conjunction   When planets or other celestial objects appear in the same, or nearly the same, direction in the sky they are said to be in conjunction. When a planet is opposite the Sun in the sky, i.e., due south at midnight, it is said to be in opposition.

Constellation   In the second century A. D., Ptolemy of Alexandria listed 48 ancient patterns in the sky representing animals, heroes and inanimate objects such as a crown, a lyre and an arrow. In modern astronomy, there are 88 officially recognized constellations in the entire sky, whose borders are defined by the International Astronomical Union (IAU). It must be remembered that other cultures outside the Western tradition use different constellations.

CPRE (Campaign to Protect Rural England)   The CPRE is a London-based national non-profit organization, with branches all over England, whose aim is the protection and enhancement of the English countryside and the promotion of sustainable land use and the management of natural resources.

Dark adaptation   In the eye, the process of dark adaptation causes dilation of the pupil and an increase in the activity of rod cells, which are able to function in low-light conditions, with a concomitant decrease in the activity of cone cells, which are largely responsible for color vision and function best in relatively bright light.

Double star   A double star is a pair of stars that appear close to each other in the sky, and these may be gravitationally connected pairs or merely line-of-sight effects. There are few double stars visible to the unaided eye, but through telescopes very many may be found.

Dwarf planet   There are five officially recognized dwarf planets in the Solar System: Pluto, Eris, Ceres, Haumea and Makemake. Former asteroid Ceres orbits in the belt of countless small rocky bodies between Mars and Jupiter, and the other four inhabit the belts of remote objects in the outer Solar System. It is likely that many more will be located in coming years. The International Astronomical Union defines a dwarf planet as an celestial body that orbits the Sun, has sufficient mass for its gravity to draw it into a more or less spherical shape (hydrostatic equilibrium), has not cleared the zone in which it moves of rival bodies and is not itself a satellite of another body.

Earth Hour   Earth Hour is a worldwide environmental event originally promoted in 2007 in Sydney, Australia, by the World Wide Fund for Nature (WWF). Every year individuals, communities, residences and businesses are encouraged to switch off all non-essential lighting for an hour.

Ecliptic   See celestial equator.

Equinox   In spring towards the end of March, and in autumn/fall towards the end of September, the Sun, moving along the ecliptic, crosses the celestial equator (in the plane of Earth's equator), making both night and day twelve hours long. These are the times of year when our planet is tilted neither towards nor away from the Sun.

Exoplanet   An exoplanet, also known as an extrasolar planet, is one that orbits a star other than the Sun. By spring 2016 more than 2000 of these bodies had been discovered. The first confirmed detection of an exoplanet was in 1992, when radio astronomer Aleksander Wolszczan of Pennsylvania State University found three planet-sized objects orbiting the pulsar PSR B1257+12, in the constellation of Virgo. The first exoplanet to be discovered orbiting a more conventional main-sequence star was 51 Pegasi b, detected in 1995 by Michel Mayor and Didier Queloz of the University of Geneva, 50 light years away in the constellation of Pegasus.

Fovea   A small central depression in the back of the retina containing cone cells, the area of sharpest vision.

Fracking   The commonly used name for the process of drilling down into the Earth and directing a high-pressure water, sand and chemical mixture into subsurface rock to release oil and gas. Fracking is now a widespread method of obtaining fossil fuels. Astronomy can be negatively affected by light pollution of the night sky if gas flaring is allowed at the wellhead and if fracking-site lighting spill is permitted.

Galaxy   A galaxy (from the Greek *galaxias* (γαλαξίας), "milky"), is a large gravitationally bound system consisting of stars in various stages of their evolution, objects orbiting stars, dust, interstellar gas and so-called dark matter. Our Milky Way galaxy of about half a trillion stars is a medium sized spiral system.

Gegenschein   In the night sky, opposite the position of the Sun and especially around midnight, a roughly oval patch of very faint nebulous light, called Gegenschein, can occasionally be made out from a very dark observing site. It is the result of sunlight scattered by interplanetary dust.

Globular cluster   A spherical agglomeration of stars, sometimes a million strong, usually orbiting the center of a galaxy at distances so far away that they are outside the main mass of galactic stars. Globular clusters are very ancient objects. The Milky Way has about 150 globular clusters, while much larger galaxies may have thousands.

Henge   In Neolithic times a common way to lay out a place of public assembly or worship was to dig out a henge, a roughly circular earthwork consisting of a ditch surrounded by an earth bank with a central flattened area. In time these developed into more complex structures using timbers and stones.

Horizon   In astronomy, the horizon is "the intersection of a horizontal plane through an observer with the celestial sphere." In other words, the line where you see the sky meeting the ground, but of course land-based observers rarely see such a thing because of local relief, trees, buildings and the like.

Light year   A light year is not a time period but a distance. It is simply the distance that light travels in one year, which is 9.4607×10^{12} km (around 9 million million km or 6 million million miles). Astronomers also use the parsec (3.26 light years), or more often the kiloparsec or megaparsec for very large distances. Within the Solar System, the astronomical unit (AU, the Sun-Earth distance of about 150 million km or 93 million miles) is often used.

Limiting magnitude   The faintest visible apparent magnitude of (usually) a star that is detectable, by the eye or through an optical instrument. In a clear dark sky, most people with normal vision can see down to magnitude +6 or at best +7. The Hubble Space Telescope can detect objects as faint as magnitude +31.

Local Group   The Local Group of galaxies has more than fifty members, including the Milky Way. The three large spirals that make up most of its mass are M31, the Andromeda Galaxy (by far the biggest), the Milky Way and M33, the Triangulum Galaxy. Most of the rest are comparatively small dwarf galaxies. The volume of the universe occupied by our Local Group has a diameter of 10 million light years. The group is itself a part of the much bigger Virgo Supercluster of galaxies (see www. atlasoftheuniverse.com/virgo.html).

Messier objects   To most astronomers, the name Charles Messier brings to mind just one thing: his catalog of 103 objects of interest in the night sky, published between 1771 and 1784. Intriguingly, this was a catalog of things to disregard, as Messier was principally hunting for comets; the nebulous and unusual objects he found, some of which had already been discovered by other astronomers, were listed to avoid confusion. Famous items in Messier's list are M1, the Crab Nebula supernova remnant; M31, the Andromeda Galaxy; M42, the Orion Nebula; M44, the Beehive star cluster; and M45, the Pleiades.

Meteor   A meteor is not an object but rather the effect of an object upon Earth's atmosphere. "Shooting stars" are the result of meteoroids (see below) being heated by collisions as they pass through the upper atmosphere, ionizing gases and creating a visible streak of light.

Meteorite   A meteorite is a survivor—a piece of rock from space that hits Earth's surface. Most meteorites are thought to originate in the Asteroid Belt and represent primitive material from the birth of the Solar System. Iron meteorites represent material that has sunk to the cores of planetesimals, which have thereafter been shattered in ancient collisions, while stony (chondritic) meteorites are crustal material.

Meteoroid   A meteoroid is a piece of debris still traveling through space. Most meteoroids are very small objects. Large meteoroids originate in collisions between bodies in the early Solar System, and some are believed to have been created in impacts upon the Moon, Mars or asteroids (for example, Vesta).

Milky Way   Our Milky Way Galaxy appears as a band of faint light in the sky, and the Western tradition of calling it the 'Milky Way' or something equally lactic in nature stems from early attempts to explain its nature, for example the Latin *via lactea* or the Greek γαλαξίας κύκλος (galaxías kýklos, the milky circle).

National Parks   In the United Kingdom, National Parks are found in England and Wales. Such areas have different names in Scotland. These are areas of relatively undeveloped terrain, but, unlike National Parks in many other countries, they usually contain settlements, in some cases of substantial size, a good deal of privately owned land and road infrastructure.

National Trust (NT)   The National Trust, or to give it its full name, the National Trust for Places of Historic Interest or Natural Beauty, is active in England, Wales and Northern Ireland. It is the largest membership organization in the United Kingdom, and one of its largest non-profit organizations. Its remit is the conservation and protection of historic buildings and landscapes. There is an independent National Trust for Scotland. The NT is also among the largest landowners in the United Kingdom, and many of its protected spaces and premises are open to the public free of charge.

Natura 2000   Natura 2000 is a network of nature protection areas within the territory of the European Union. It consists of Special Protection Areas (SPAs), Special Areas of Conservation (SACs) and Marine Protected Areas (MPAs), all designated under EU environmental directives.

Ordnance Survey   Ordnance Survey (OS, since 2015 Ordnance Survey Ltd) is Great Britain's national mapping agency. It is one of the world's largest cartographers and distributors of maps.

Perigee When the Moon is at its closest point to Earth in its orbit, it is said to be at perigee.

Perseid The Perseids are one of the most reliable meteor streams, and consist of debris left behind by Comet Swift-Tuttle. Like other meteor 'showers' (an ill-chosen term since they almost never fall like rain!), the Perseids are named after Perseus, the constellation containing their radiant, which is the point in the sky from which they appear to emanate. They are seen at their best around August 11–13 every year, and one of their more attractive alternative names in predominantly Roman Catholic countries is the "Tears of Saint Lawrence," since they fall around the saint's feast day.

Phases of the Moon As the Moon revolves around Earth it appears to change shape in the sky, developing from invisibility when in line with the Sun and not eclipsing it (new Moon) into a crescent (Latin *crescens*, growing, and in French *croissant*. Have you ever wondered where the name of those Moon-shaped buns comes from?) A few days on, and we see the right-hand side of the Moon illuminated (first quarter). Although we then see half of the face that the Moon always presents to us, it's called a quarter because it's a quarter of the way through its cycle of phases. Full Moon occurs when the Earth-facing side is fully lit by the Sun, then the pattern is reversed as the Moon wanes, through last quarter and back to new Moon as the lunar month ends.

A little thought will confirm that there is indeed no such thing as moonlight, simply the light of the Sun reflected from the Moon onto Earth. A futuristic astronaut standing on the Moon during the fourteen-day-long lunar night would likewise see Earth 'shining' down upon her—sunlight reflected from Earth, or Earthshine as astronomers know it. This phenomenon is easily seen from Earth by looking at the Moon in a dark sky when it is at crescent phase, with the night-time part of its disk faintly visible. Use binoculars to enhance the effect.

Planetarium A building, usually dome-shaped, where an artificial night sky is projected onto the ceiling. The rotation of Earth, the phases of the Moon, meteors and the motions of the planets can be simulated. There are also mobile inflatable planetaria. The word *planetarium* is becoming increasingly used for astronomy apps and online interactive night sky maps.

Planetesimal Small rocky body formed during the early evolution of the Solar System, from accretions of dust, rock, and icy materials. The early planetesimals varied in size from just a few meters to hundreds of kilometers, and collisions between them led to the formation of both larger asteroidal bodies and small meteoritic fragments.

Ramsar Convention The Convention on Wetlands of International Importance, especially as waterfowl habitat (Ramsar Convention or Wetlands Convention) was adopted in Ramsar, Iran, in February 1971 and came into force in December 1975. The Convention covers all aspects of wetland conservation. The convention's mission is "the conservation and wise use of all wetlands through local and national actions and international cooperation, as a contribution towards achieving sustainable development throughout the world."

Red giant There are several red giant stars easily visible in the night sky. Examples are Betelgeuse and Aldebaran, in Orion and Taurus, respectively. Red giants are stars in a late stage of their evolution, having swelled to enormous size because of chemical reactions in their cores. A few billion years hence, our own Sun will become a red giant star, engulfing Mercury and Venus as it expands, and probably searing Earth in the process.

Reflector The reflecting telescope (or reflector), invented by Isaac Newton in 1668, uses a single or sometimes two curved mirrors to reflect light and form an image as seen through an eyepiece. This was a partial solution to the false colors so prevalent in refracting telescopes (see below) of the time, and also allowed its user to observe sky objects towards the zenith without having to crouch down low, as the eyepiece is set at right angles to the telescope tube and can be used by an observer standing up.

Refractor The refracting telescope (or refractor) uses a lens or combination of lenses at its skyward end to form an image, viewed through an eyepiece set normally at the other end. This kind of telescope was invented around 1600 by a Dutch optician, Hans Lippershey.

Rods and cones Cells in the retina of the eye. Rods are cylindrical cells containing rhodopsin ('visual purple'), and are sensitive to dim light but not to color. Cones are conical cells that are sensitive to color and bright light. The process of dark adaptation involves the rods taking over visual duties from the cones. Interestingly, there are no rods in the center of the fovea (see above), which explains the astronomer's 'averted vision' trick (objects appearing more distinct if you look slightly to one side of them).

Satellite The general meaning of the word *satellite* is "anything that revolves around something else," for example a communications satellite orbiting Earth or the Galilean moons orbiting Jupiter. Hundreds of artificial satellites currently orbit Earth, or are in relatively fixed positions (geosynchronous orbits) above it. They are used for a great number of different purposes, among them weather forecasting, television transmission, radio and Internet communications, navigation (GPS) and climate research.

Seconds of arc Angular measurements of astronomical objects are expressed in degrees, which are divided into arcminutes (minutes of arc) or arcseconds (seconds of arc). A degree is divided into 60 arcminutes, and an arcminute is divided into 60 arcseconds. The sky apparently rotates due to the rotation of Earth at about 15 degrees an hour. The diameter of the Sun, and the Moon coincidentally, is approximately half a degree.

Spotting 'scope A spotting 'scope is a small portable telescope with added optics to present an erect (right-way-up) image and is normally used for terrestrial objects; it is often used, though, for astronomy.

Standing stones/stone circles In various parts of the world as far apart as Israel, Hong Kong, West Africa and western Europe, there are ancient rings or ellipses of stones, most likely erected for ceremonial and religious purposes. The greatest densities of stone circles occur on the landscapes of the British Isles and Brittany in France. In these two areas there are more than a thousand examples, nearly all dating from the Late Neolithic or Early Bronze ages. Many have been shown to have some astronomical significance, the most famous being the alignment of the main axis of Stonehenge with the rising midsummer Sun.

Star party Star parties are gatherings of observers, often members of an astronomical group with invited members of the public, and usually in a relatively dark place. These sessions are becoming more and more popular, and are an indication of the public's growing interest in the oldest science.

Supernova Some massive stars will collapse rapidly at the end of their lives, triggering a catastrophic explosion that greatly increases their brightness and ejects most of their mass. Within these mighty furnaces, many of the heaviest and rarest elements are created. It's an interesting fact that we pay a lot for gold, platinum and other rare metals because supernovae occur infrequently—possibly about one in any Milky Way-sized galaxy per century.

World Heritage site All over the world, the World Heritage Convention (part of UNESCO) designates and seeks to protect sites, areas or structures of 'outstanding international importance' as World Heritage sites. It is a sad fact that the night sky at these sites does not appear to receive this protection, and in certain places is lost to floodlighting and, in the case of the Stonehenge area, to skyglow from a nearby town and a golf driving range.

Zodiacal band and zodiacal light The zodiacal light is a faint, cone-shaped pale glow along the ecliptic in the night sky that appears to extend upwards from the horizon where the Sun has set or will rise. It is best viewed around the spring equinox as night falls, or before sunrise in a dark sky around the autumn equinox. The effect is caused by sunlight scattering from dust in space in the near Solar System, and light pollution has made it a thing of the past in many developed places. The zodiacal light decreases in brightness further from the Sun's position in the sky. From *very* dark sites it can appear as a continuous zodiacal band along the ecliptic. In fact, the zodiacal light would extend across the whole sky in the absence of other light sources.

Appendix I

Glossary of Light Pollution Terms

Following are words and phrases it can be useful to know when discussing light pollution issues.

Ambient light. The total light level or effect, or amount of light perceived, in one's surroundings.

Baffle A plate inserted within or just outside a luminaire to shield the light from direct view.

Ballast Electrical devices used in conjunction with a discharge lamp to start and control it.

Candela (cd) or standard candle The SI unit of luminous intensity.

CfDS The British Astronomical Association's Commission for Dark Skies. A non-profit organization working towards the optimum night sky for all, wherever they live, through proper control of lighting.

CFL Compact fluorescent lamp.

CIE The Commission Internationale de l'Eclairage (International Lighting Commission), based in Vienna. Europe's premier authority on lighting matters.

Color rendering/rendition The perceived effect on objects of different colors of lights of different types.

Color temperature This term describes the actual color of the light source itself, as opposed to that of the light issuing from it.

© Springer International Publishing Switzerland 2016
B. Mizon, *Finding a Million-Star Hotel*, The Patrick Moore Practical Astronomy Series,
DOI 10.1007/978-3-319-33855-2

Column   The post upon which a lamp is mounted.

Dark sky ordinance   Many towns, counties and even regions across the world have enacted local regulations requiring polluting lighting fixtures that waste light in various directions, and especially into the night sky, to be replaced. Objectives of such ordinances are the prevention of intrusive light, the reduction of skyglow, glare and energy waste and the promotion of good lighting practices.

Disability glare (veiling luminance)   Glare causing reduced visual performance.

Discomfort glare   Glare producing discomfort or annoyance without necessarily interfering with visual performance.

FCO   Full cut-off, or a lamp with a flat glass panel beneath which, when mounted horizontally, emits no light above the horizontal. See also SCO.

Fluorescent   A long-life, relatively cheap whitish light source based on a gas discharge process, where electrons pass through a tube and interact with a phosphor coating.

Flux   Luminous flux is the rate of flow of particles of light energy, measured in watts or ergs/sec.

Full cut-off   See FCO.

Good Lighting Award   Several anti-light-pollution movements give awards for good lighting practices. The twin aims are to reward those who have chosen environmentally responsible lighting and installed it in such a way as to confine it to the place to be lit, and to create publicity about light pollution to raise public awareness of its deleterious effects.

IDA   The International Dark-Sky Association, based in Tucson, Arizona. The world's first anti-light-pollution organization and the leading organization combating light pollution worldwide. Founded in the United States in 1988, the IDA now has sections in many other countries across the world.

IDA Europe   The International Dark-Sky Association has branches in several European countries, operating as IDA Europe. The British Astronomical Association's Commission for Dark Skies is an affiliate of the IDA and works closely with it.

IESNA/IES   The Illuminating Engineering Society of North America. The professional guidance body for lighting engineers in the United States.

ILP   The Institution of Lighting Professionals. The UK's professional guidance body for lighting engineers.

Incandescent   Describes a light source based on electricity passing through a thin filament (usually tungsten) that glows brightly.

Intrusive light   See light trespass.

ISTIL Istituto di Scienza e Tecnologia dell'Inquinamento Luminoso (Light Pollution Science and Technology Institute) is a non-profit Italian organization with the aim of developing and promoting scientific research on light pollution, and developing and spreading technologies and methods to limit light pollution and its adverse effects on the night-time environment.

kWh Kilowatt-hour; unit of energy equal to the work done by 1000 watts of power acting for one hour.

LED Light-emitting diode. Currently, the main source of white light on our roads and likely to become the source of choice for just about all lighting in the coming decades.

Light spill The emission of light outside the premises in which the lighting is supposed to illuminate.

Light trespass (intrusive light) Troublesome light entering areas or premises outside the boundary of the premises to be illuminated. UK campaigners against light waste tend not to use the term 'trespass,' which has a specific meaning in law. It is not normally the intention of the owners of intrusive lights to cause a problem. They simply do not realize that they are being intrusive.

Lumen (lm) The SI unit of luminous flux, being the flux emitted in a solid angle of 1 steradian by a point source with uniform intensity of 1 candela (q.v.).

Luminaire A word not found in many dictionaries but widely used in the lighting community to denote the lamp and its surrounding casing and optics.

Lux (lx) The SI unit of illumination, being a luminous flux of 1 lumen (q.v.) per square meter. The value for the full Moon is about 0.2 to 0.3 lux.

Melatonin A hormone (N-acetyl-5-methoxytryptamine) secreted during the hours of darkness by the pineal gland.

Obtrusive light Light emitted where it is not needed, causing nuisance or environmental degradation.

Photometry The measurement of the level and distribution of light.

Reflectance The amount of light reflected by a given surface (the ratio of the reflected flux to the incident flux).

Reflectivity The ability of a surface to reflect radiation (technically, equal to the reflectance of a layer of material sufficiently thick for the reflectance not to depend on the thickness).

SCO Semi-cut-off: a lamp type that has a shallow bowl beneath, and emits little or no light skywards.

SON Another name for high-pressure sodium light sources.

SOX Another name for low-pressure sodium light sources.

Skybeam, sky beam   A concentrated beam of light sent into the sky deliberately, usually for the purposes of advertising (often erroneously called a 'laser').

Sky glow, skyglow   Unwanted light emitted into the night sky from poorly aimed lamps. Skyglow in rural areas emanates from sometimes quite distant light sources. Skyglow over urban areas is often referred to as a 'light dome.'

Sky quality meter (SQM)   An SQM measures sky brightness in magnitudes per square arcsecond. Designed by Dr. Doug Welch and Anthony Tekatch, this hand-held instrument gives a reading when pointed at the sky and activated. Really dark skies will score around 22 and 23, and very poor, polluted urban skies will score as low as 17. Users must be careful to avoid local light sources such as car interior lights, phones and the like. Also, the presence of nearby trees can affect the accuracy of readings.

Stray light   See light spill.

Street furniture   All manufactured items commonly seen along roadsides, for example lighting columns, traffic signs and telephone poles.

Veiling luminance   See disability glare.

Visibility   Clarity of vision; how well we see something. The purpose of a good light should be to increase visibility—to reveal and not conceal. See disability glare.

UWLR, ULR, Upward flux   The abbreviations stand for Upward (Waste) Light Ratio. All these terms refer to the relative amount of the light emitted above the horizontal if lamps are not appropriately shielded.

Index

© Springer International Publishing Switzerland 2016   319
B. Mizon, *Finding a Million-Star Hotel,* The Patrick Moore Practical Astronomy Series,
DOI 10.1007/978-3-319-33855-2